PICK-YOUR-OWN FARMING

Free Yourself from Farmers Markets and Join the Agritourism Revolution

Megan Neubauer

Acres U.S.A.
Viroqua, Wisconsin U.S.A.

About Acres U.S.A.

Founded in 1971 by Charles Walters, Acres U.S.A. emerged from the need to promote ecological farming practices in a time when industrial agriculture was heavily reliant on synthetic fertilizers and pesticides. Inspired by figures like Rachel Carson and Dr. William Albrecht, Walters used the magazine, and later books and conferences, to advocate for sustainable agriculture that prioritized soil health and natural processes. Acres U.S.A. provided a platform for these ideas and helped to popularize alternative methods like cover cropping and integrated livestock management.

Though the agricultural landscape still relies heavily on conventional methods, Acres U.S.A. has been instrumental in the growing movement towards regenerative agriculture. By disseminating knowledge and supporting eco-conscious farmers, the company continues to champion sustainable practices through its publications, conferences, and online resources, contributing to a shift towards a more environmentally sound approach to farming.

Find Out More About Acres U.S.A.

Pick-Your-Own Farming: Free Yourself from Farmers Markets and Join the Agritourism Revolution

Cover photography credits: Christie Connell, Megan Neubauer
Interior illustration credits: Kendyll Harden

Acres U.S.A.
PO Box 351
Viroqua, Wisconsin 54665 U.S.A.
1 (970) 392-4464 • info@acresusa.com • www.acresusa.com

Printed in the United States of America
Megan Neubauer, 1982-
Pick-Your-Own Farming / Megan Neubauer. Greeley, CO, ACRES U.S.A., 2023. 170 pp.
ISBN 978-1-60173-1845

Library of Congress Cataloging-in-Publication Data
Name: Neubauer, Megan 1982 – author
Title: Pick-Your-Own Farming: Free Yourself from Farmers Markets and Join the Agritourism Revolution
Description: Greeley, Colorado: AcresUSA, 2023

Library of Congress Control Number (LCCN): 2023948108
1. Agriculture 2. Economics

ADDENDUM TO PICK-YOUR-OWN FARMING:

This book was published in 2023, and some product names mentioned in the book have changed since then.

One product, PhytoMast®, is the officially registered name for what was called Phyto-Mast in this book. The product is an herbal blend of thymol (antiseptic) balanced with angelica, licorice, wintergreen, olive oil, and vitamins A, D, and E. It is packaged in aseptic infusion tubes (with alcohol pads) to be used as intramammary infusion for healthy cows at dry-off. Peer-reviewed, published articles in studies done with dairy cows and goats have shown its utility and safety as a dry-off product. PhytoMast® is labelled specifically with commercial Grade A inspected dairy farms in mind so the farm can meet the regulations of the Pasteurized Milk Ordinance (PMO) item 15r (Storage of Drugs and Chemicals).

Along the way, I enhanced the formula and added cinnamon bark oil (CBO) due to many published studies in the literature demonstrating the destructive effects of CBO upon biofilms that various pathogenic germs form to protect themselves. To keep these similar yet different products separate, the name UdderWell is used for the formula with CBO. UdderWell should not be on the shelf of Grade A inspected dairy farms. In other words, UdderWell is labelled for those who are raising animals for themselves and not for commercial use. The difference between the two products, other than the presence of CBO, is strictly a labeling issue and not a quality or concentration difference.

UdderWell, GetWell, BreatheWell, EatWell, LivWell, BreedWell, and FeelWell are available at www.reverencefarms.com

Pages	New product names
331	**UdderWell/ PhytoMast®** for step (3)
206, 220, 238, 243–245, 269, 290, 373	**GetWell** for herbal antibiotic tincture; for herbal antibiotic mix; for Phytobiotic; for tincture of garlic, echinacea, goldenseal, wild indigo, and barberry
210, 218, 221, 223, 225, 233, 373	**EatWell** for tincture of ginger, gentian, nux vomica, and fennel; for Digestive Ø; for Phyto-Gest
210, 214, 217, 373	**LivWell** for tincture of red root, celandine, milk thistle, dandelion, and oregon grape root; for Phytonic
373	**FeelWell** for Phyto-gesic; for herbal pain formula tincture
284, 289–292, 373	**BreedWell** tincture / **HeatSeek** powder for "Spectra 305"
232, 238, 239, 243–245, 302, 327	**AmpliMune®** for ImmunoBoost; see www.novavive.ca
232, 238, 244, 245	**BoviSera®** for Bo-Bac-2X/ Quatracon/ BoviSera; see https://colorado-serum-com.3dcartstores.com

For the brave souls who dedicate their lives
to feeding and healing the planet for my daughter
and all of our children.

Table of Contents

Acknowledgements

First thanks are due to my little family, Allan and Quinn, for supporting me throughout the long journey of building a farm and writing this book. I'm especially grateful you take such good care of each other every summer when Mom's busy at the farm. My Dad has provided immeasurable support for this project not only as a great Dad but also my business partner, and every single aspect of the enterprise on these pages was conceived by us together. My Mom has cared for our daughter as a third parent since birth, and without her support the farm could never have become what it is today. Last but not least, thank you to my soft little writing buddy who couldn't stick around long enough to see this in print (he probably wouldn't have been impressed, anyway).

We are fortunate to belong to an incredible community of North Texas farmers, advocates, and friends. I'm deeply grateful for the support of everyone here in McKinney, especially Rick Wells and everyone at Wells Group, the Seed Project Foundation, and McKinney Roots. To Lynne Sipiora, Pam Walker, Scott Elliott, and Bruce Miller, your personal encouragement means more than I can say. Thank you to Judith McGeary and her team at Farm and Ranch Freedom Alliance for their tireless advocacy to create a more fair and prosperous local food system for small producers like us. Maria Whitworth and Nanci Taylor also do amazing work with *Edible Dallas & Fort Worth* to show our faces and tell our stories. To my hype team of friends here at home and around the world, thank you for keeping my spirits high.

Thank you to Lydia Lazar and the team at Acres USA for helping us share our journey, and to Ryan Slabaugh for inviting us into the fold. I must also thank Carolyn Roarke, Ben Hartman, Natalie Wallace, and my sister Marlene Neubauer for their invaluable guidance in getting started. To my wonderful illustrator Kendyll, I am grateful for your sheer delight in growing things. Thank you for your dedication to this project.

I want to thank our extraordinary team at Pure Land Farm, especially Teri, Deanna, Baylee, and Marylinda, for shaping the farm into what you see today. I say it often, but it's never often enough; we couldn't do this without you.

Finally, to the 25,000 pickers we've hosted over the last six years, thank you for choosing to spend a day in the dirt with us.

Introduction

Dad and I started Pure Land Farm in 2012. We were looking for a piece of land to purchase in the rash ingredients and love of cooking in me from an early age that only grew over time, so we hatched the idea of growing food as a business. We settled on a beautiful 28-acre property in McKinney, a long, narrow plot that gently slopes down to Honey Creek. Dad had recently retired from a long career in petroleum engineering (we call this second career in farming his "atonement period") and I resigned from my job in Pediatric Oncology research at UT Southwestern Medical Center to give this plan a shot.

From the beginning, we committed to using only organic, regenerative farming techniques, producing as little waste as possible, and operating as a twosome, without relying on free volunteer or family labor. We took out a loan on a tractor, signed up for a booth at a local farmers market, and got to work. With no experience in agriculture beyond backyard gardening between us, the learning curve was steep. But we gave ourselves some grace and started meeting other farmers and making friends in the local food scene. One engineer plus one scientist has the promise of making one *decent* farmer, so it didn't feel like we were starting from ground zero. I was fortunate that my husband's career provided financial security and health insurance for my family, which gave us a cushion of time and stability to figure this farming thing out. But from the beginning, we weren't interested in farming as a hobby. Like anyone starting any business, we wanted to make money. The income from our sales at the market was quickly spent on supplies and tractor payments, but I always expected success to be *just around the corner*. We were working so hard; surely we would actually get to take money home one day… right?

Every year it was more difficult to justify staying in business. Saturday after Saturday, even the best growers at the farmers market would only sell a fraction of what they brought in, loading perishable product back into their trucks and starting the hustle to find buyers. A farmer's retail (or highest) prices are usually only fetched at the farmers market, so any unsold produce immediately drops in value and quality after spending hours outside in the Texas heat.

After a couple of years, it dawned on us that we were capable of growing much more than we could sell. To pick, prepare, market (meaning physically take to the buyer for sale, such as going to the farmers market or delivering to a restaurant or distributor), and sell everything we *could* grow, we would have to not only work much harder but probably hire help and make expensive upgrades in infrastructure like walk-in coolers, further reducing our meager profit. Working harder isn't particularly appealing when you're already bunching carrots at 11 o'clock on Friday night for the next morning's market. We also took pride in operating our farm as a pair and wanted to stay as small as possible provided we could make it work financially. If we stayed the course and only grew as much as the two of us could pick, prep, and market, even streamlined to the maximum efficiency, the revenue wouldn't be enough to pay minimum wage for our labor, let alone provide a decent living. Simplifying by selling in bulk feels like giving your hard work away when you hear what wholesale customers are willing to pay, and it didn't cut out that much work anyway. Everything still had to be picked, prepped by washing, sorting, bunching, or bagging, and delivered.

When the entire revenue for a business is derived from the sales of essentially one product - in our case, fresh produce – one key step on the critical path to make more money is to make more product. We *had* to grow more food. And, with a few years of experience under our belts, we felt confident about being able to do it. The question then became, how were we going to sell it all without killing ourselves?

In 2018, we changed Pure Land's business model from a market garden to a pick-your-own, effectively reallocating the picking, prepping, and marketing of our product to the consumer. To be precise, we incorporated the picking, prepping, and marketing *into the product*, and started charging a premium for it. By replacing all other direct-to-consumer sales with agritourism, we've finally planted ourselves firmly in that rarest class of small farms: profitable.

Back in our farmers market days, one question we heard often from across the table was, "Can I come out and see your farm? I'd love to show my kids where food comes from." We obliged, when we had time, and it was refreshing to witness the joy and fun of seeing a real farm for the first time through a visitor's eyes. Showing a five-year-old how to properly grip the greens to pull an eight-inch carrot out of the soil gets a bigger reaction than any magic trick I've ever seen. Growing food *is* kind of a miracle, something easy to forget after a few years of squash bugs and sore backs.

It started with the berries. In 2015, we planted a small patch of thornless blackberries to see how they would fare in our black clay soil. They tasted great, commanded a good price, and produced well enough that in a moment of madness the following year, we expanded into a field of 800 canes. Then the thought occurred to us: *how are we going to pick all of these?* We researched pick-your-own berry farms and discovered we'd be the closest one to Dallas, a massive market, so we decided to open the blackberry patch to pickers and continue taking everything else to the farmers market.

Immediately upon opening the blackberry picking season that first year, visitors wanted to know when they would get to pick the tomatoes over there… and *wow, are those onions?…* and *oh I see carrots over there, can you show my kids how they grow?* The excitement over the blackberries was expected, but the desire to pick everything else took us by surprise. There are no other farms in our area that operate a diversified pick-your-own, so we had no one to ask for advice or model after, and we could find no resources on the subject. Honestly, I wanted to buy the book in your hands, not write it. Nevertheless, the clear interest from customers convinced us to forge ahead. Over the following winter, we took a leap of faith; we rearranged the planting plan, relinquished our spot at a popular farmers market, and crossed our fingers that people would come.

To our great relief, they did. That first summer, we hosted about 2,000 visitors and learned a lot very quickly about operating a pick-your-own. It was promising enough to convince us to quadruple the size of our blackberry field. After fine-tuning the nuts and bolts, including the addition of an online reservation system to ensure the right number of people for the yield of produce ready to harvest, we hosted 5,000 the second year. We hosted 7,500 visitors the following year and are limited solely by our production capability. According to our website traffic data, the number of people seeking reservations to visit the farm has spiked at times up to ten-fold higher than we could host.

It's not magic. It's still a lot of hard work, and the job is not for everyone. But in our experience, operating a farm as a pick-your-own is less stressful than growing for market and the potential profitability is much higher. Our aim with this book is to encourage more small farms to fully transition to or at least incorporate some agritourism in their operations. If we can help our fellow farmers achieve profitability, we can stop them from closing and even encourage new farms and growth. To be clear, this is not a book of gardening advice that will teach you how to grow more; this is a book that will show you a way to *sell* more, with agritourism. Once your operation has created more demand than you can produce, you'll need to order the many books that teach you how to grow more.

The dirty secret among market gardeners is that most of their businesses aren't profitable. Over the last few generations farmers have been told to "get big or get out," while government policy and consolidation in the food industry have overwhelmingly supported large-scale commercial agriculture. Small farmers have struggled to compete and survive. To quote regenerative agriculture pioneer and advocate of the pick-your-own concept Dr. Booker T. Whatley back in 1985,

> "Although the USDA and the land grant college agriculture

> departments are pushing diversification, [they are] relying on traditional marketing outlets like farmers markets and cooperatives. That just won't work. Small farmers simply can't afford to pick, grade, wash, package and haul their produce maybe 100 miles or more to market, yet that's exactly what the so-called experts are telling them to do."

Many new and aspiring farmers are faced with accepting that an ostensibly bucolic life spent turning the earth is its own reward; allow me to tell you from years of personal experience, *it is not.* Most new farms that fail within a few years do so not for lack of effort or passion, but because they can't make enough money to pay themselves. Like any other business, farming has to be profitable to be sustainable, and farmers deserve to pocket cold hard cash for their labor.

In a 2015 article for *Salon*, Jaclyn Moyer described the struggle to make a living on her small farm in Northern California. Curious, she set out to see how many other farmers were in similarly dire financial straits and set three criteria for what she considered "making a living": the farmer had to make at least minimum wage for their area, use no unpaid workers or volunteers, and rely on no income from outside of the farm. She did not find a single farmer that met all three. Less than a year later, she sold her farm.

Many circumstances contribute to this struggle. For one, being unable to make enough money from farming causes many to rely on outside sources of income to make ends meet, and if there's one thing guaranteed to prevent a person from making more money as a farmer, it's having a second or third job that keeps them away from farming.

Another factor that keeps small farms from turning a profit is lack of startup capital. Many fledgling farms purchase the bare minimum to get by, relying on used (read: old, broken) equipment, cobbling contrivances together from the junk pile and forgoing some necessary inputs altogether while just hoping for the best. After all, it's much cheaper to cross your fingers and toes when the temperatures dip below freezing than to buy frost cloth, especially when you may not even need it. Some years, you get lucky! But one year, you'll find some frost cloth on sale and snap it up, and then use it for several seasons and wonder how you ever tried to get by without it. It takes time to identify and afford the right tools and equipment to manage a farm, and the quicker one can acquire them by starting off profitable, the better. A farm that can't buy the things it needs to operate effectively will never reach its potential.

This compounding in effectiveness over time applies to farming skills, too; to get better at growing, a farm must be able to stay in business long enough that the farmer can *learn how to grow.* Even a farm with a fortune in startup capital is not guaranteed success. It takes time to learn how to farm, and there's no shortcut. You can watch all the YouTube in the world but it's no replacement for experience.

Complicating matters is the natural complexity of a living biological system where everything in it is dependent on everything else, and everything is constantly changing. If a novice baker burns a loaf of bread, they can toss it out and whip up another in hours. But if I lose my tomato crop, I'm out an entire year before I get to try again. Eliot Coleman pioneered modern small-scale sustainable agriculture on his farm in Maine which now pulls in about $100,000 per acre on organic fruits and vegetables. Before you start seeing dollar signs, consider that Eliot Coleman started Four Seasons Farm *over fifty years ago*. How long will it take *you* to become skillful enough to grow $100,000 an acre? To quote a young farmer I spoke with recently near the end of his first season, "I knew growing stuff wouldn't be *easy* – but does it have to be *so hard?*"

To be sure, not every person seeking fresh produce has the luxury of time and money to make their way out to our farm to pick vegetables every week. There are deep, troubling issues of inequality and food sovereignty in this country, but I cringe when the burden of mitigating these is placed on farmers. We can argue about the possible ways to make fresh, organic food more affordable, a worthy endeavor to be sure, but none of them work if the farmers can't pay their bills. Nonprofit farms are on the rise and doing incredible work towards food and social justice, but businesses are the engines of our economy. A profitable business is *truly sustainable*. My hope is that when there are many more small farms like ours thriving, a swell of support for local agriculture will prime the environment for building a more just, diverse, robust food system.

An illustration of the fragility of our current food system came to light during the COVID-19 crisis. Bare shelves at the grocery stores caused panic among folks that don't frequent farmers markets or outlets stocked by local producers. Many farmers I know had record sales for the first couple of months, until the supply chain evened out and customers reverted to their old buying habits. Small, family-run farms produce 70% of the food consumed globally; if we are to feed the world and build resilience in our food system, we need every small farm to succeed.

In 1987, Dr. Booker T. Whatley released his guidebook *How to Make $100,000 Farming 25 Acres*, published by Rodale Press. In the book, Whatley asserted that a you-pick model requiring customers to pay for access through a membership program is essential to a small farms' success. His strategy is summarized in what he called Dr. Whatley's Ten Commandments.

Thy small farm shalt:

1. Provide year-round, daily cash flow.
2. Be a pick-your-own operation.
3. Have a guaranteed market with a Clientele Membership Club.
4. Provide year-round, full-time employment.
5. Be located on a hard-surfaced road within a radius of 40 miles of a population center of at least 50,000, with well-drained soil and an excellent source of water.

6. Produce only what the clients demand - and nothing else!
7. Shun middlemen and middlewomen like the plague, for they are a curse upon thee.
8. Consist of compatible, complementary crop components that earn a minimum of $3,000 per acre annually.
9. Be 'weatherproof,' at least as far as possible with both drip and sprinkler irrigation.
10. Be covered by a minimum of $250,000 worth ($1 million is better) of liability insurance.

With the exception of pursuing year-round production (because we choose not to), every concept in this decades-old list is organically represented in our own farming and business philosophy. After retiring from his career in academia, Whatley traveled the world to speak to farmers and wrote extensively on his ideas. He is largely credited with creating the niche in which small farms like ours can succeed today.

While we have clearly not invented the concept of the pick-your-own farm, there is a lack of practical and current information available for farmers in its execution, from marketing and finding customers, to honing your crop mix and maximizing production, to ringing up sales and keeping the books. We have strategized a growing plan, streamlined the operational kinks, and employed simple modern technology to our advantage. The only staff we've taken on are greeters and cashiers during the picking season; we still manage all the growing as a twosome and have found our sweet spot at two acres of blackberries and two acres of annual fruits and vegetables. It simply isn't necessary to go big or go home. What we do at Pure Land can work on any scale, as small or as big as you want to be. And it isn't necessary to live as an ascetic or spend every waking moment working at the farm, either.

Our farm's evolution into agritourism has had consequences that extend far beyond the bottom line. The regenerative practices in use on small farms like ours, such as minimal tillage and planting cover crops, can help to blunt the effects of climate change. On top of that, we get to loudly preach our farmy, earth-lovin' gospel to the hundreds, sometimes thousands of people we host every week. We don't *tell* them what we do from behind a table at the farmers market, we *show* them. We are cultivating a powerful connection between nature and our bodies, a bond that's often lapsed or was never established in our complicated modern lives. When a guest visits our farm, their senses become fully engaged. They feel the hot sun on their backs, discover the prickles on a cucumber that's never been touched, revel in the heady scent of tomato leaves and moist earth. Inevitably, after an hour of playing farmer, they realize, *this is hard work.* Many of our guests become inspired to spend more time outdoors or start gardens of their own (don't worry - they always come back). More than a few linger at the crest of the hill, gazing out over our tidy rows of green, and softly say, "I wish *I could do this.*"

You can. Frankly, we *need* you to. Perhaps you have no experience in agriculture but dream of escaping the rat race to spend your days outside doing honest, meaningful work. We'd like to see new farmers break ground through agritourism instead of grinding away in a system that no longer works for the small local farmer, where too many burn out and throw in the towel after a couple of years. There's a better way to operate a small farm, one where the chosen few with the will and passion to work their hineys off to nourish their communities can pay their mortgages, too.

It is in that spirit that we offer you this book, and hope you'll take the leap into agritourism with us.

Part 1

AGRITOURISM IN THEORY

Chapter 1

THE ADVANTAGES OF AGRITOURISM

Before converting to a pick-your-own, Pure Land Farm was a market garden for five years. This type of farm is typically a small, diversified operation where the farmer brings their harvest to local sales outlets in the community. Overwhelmingly, market gardens rely on direct-to-consumer sales; in 2011, the USDA's Economic Research Service reported that two thirds of these small producers earned at least 75 percent of their gross sales through direct-to-consumer transactions, the most common venue for which is the farmers market.

It's no wonder most new market gardeners head straight for the market, as it has some appealing qualities. The customers are already there, the barrier to entry is low (usually a reasonable weekly fee for the booth space), and the regulations to sell unprocessed, self-grown produce are not complicated. Just grow some stuff and show up. For a new farmer, it might be worth starting out at a farmers market for a short time before launching into pick-your-own to get to know the local food community. There is clear value in marketing, as shoppers at the farmers market are many of the same people who will make the effort to come to your farm, but there's another benefit too - farmers markets are full of farmers. It's well worth taking the time to make friends with fellow local growers. When we started out, I was amazed at how generous our peers were when we asked for advice; there are no proprietary techniques, no secret pest control regimens, no seed hoarding to speak of among most small farmers growing food for their communities. If one happened to crack the code on a certain crop, they'd lay out exactly what they did, step by step, taking pains to explain where they made mistakes so that someone else might avoid them.

Those farmers already knew what it took us a couple years to realize; when one small farm succeeds, all small farms succeed. We're not really in competition with one another. Shoppers at a farmers market expect to see fresh produce above all else, and attractive, high-quality offerings can entice the shopper who used to only buy eggs and vegetables to explore local cheese and beef too. A customer who becomes a convert, a True Local Food Believer, evangelizes to their community and brings others into the fold. When you spend a dollar with a locally owned business, seventy cents of it stays in your community. That money doesn't just disappear into the internet, like most of our purchases these days; it ripples through your local economy. The people who own those small businesses live in your neighborhoods. When you shop at their establishments, they can pay their mortgages and feed their families. They can donate to fundraisers at the schools your children attend together and support the nonprofits that serve members of your community in need. And they can do all of this while they shop at other locally owned businesses, compounding the impact! Win, win, win.

The Limitations of Farmers Markets as a Primary Sales Outlet

Although they've existed in their modern form since the 1600s, America's first boom of farmers markets hit in the 1970s. Healthy eating went from niche to trendy as Americans started taking more interest in sourcing and cooking with fresh produce. The movement has only grown in popularity since; between 1994 and 2019, the number of registered farmers markets in the US more than quadrupled.

With the increase in demand for high-quality local produce, small farms also increased in number. Add to this the promise of wholesome country living and sunny days spent in service of feeding our communities and it's easy to understand why many are called to a life of farming. In the 2017 USDA Census of Agriculture, the number of small farms (defined as one to nine acres) in operation came in at 273,000.

Initially, the supply and demand seem to line up perfectly. Data suggest there is ever-growing demand for high-quality local produce and that there are plenty of eager small farmers trying to meet it. But it's more complicated than that. In a recent seven-year study in Oregon, nearly half of the 62 new farmers markets that opened had closed by the end of the study, almost a quarter of those within their first year of operation. There are lots of reasons farmers markets fail. They can be poorly run by unskilled managers, they're often unable to attract high-quality or in-demand vendors, and many areas are simply too saturated with markets to draw sufficient traffic to support them all. We see this phenomenon in our neck of the woods every year. Despite one or more existing markets mere miles away, there's always someone starting a new market. Everybody wants their very own farmers market in their neighborhood, and on the surface, it seems plausible that creating more sales venues in this way might enable farmers to sell more. Unfortunately adding more markets often has a deleterious effect on the sales at each one, and even good new markets can take years to gain the kind of following that translates into consistent sales.

Over our first five years of farming, we sold at about half a dozen different farmers markets in the Dallas area and our experiences were varied. Good markets have reliable, predictable traffic, an appealing mix of artisans, and a fair representation of what is locally available by season. The best markets support their vendors and create a pleasurable shopping experience for the customers. The very best often have one simple thing in common, the lack of which has tanked many a Saturday of sales in our market days: a roof. The worst markets tend to be inflexible with vendors and lax about their rules, leading to low engagement and ultimately, vendor resentment. Some markets are certainly managed better than others, but there are limitations even a well-run farmers market cannot overcome.

In the beginning, we sold out at every market and were rather proud of ourselves. It didn't take long to realize that was because we weren't very good growers. It's not hard to sell out when you aren't putting much on the table. Over time, as we became not-terrible growers, the problem turned into selling everything we grew without killing ourselves selling at multiple markets and delivering produce all over town, which added up to a lot of time spent *not farming*.

Pure Land Farm produce for sale at farmers markets.

Before I tear them to shreds, I want to say unequivocally that farmers markets are not all bad, and we are not suggesting they be abolished. Farmers markets benefit communities in myriad ways, and many of their flaws are not in their control to change. To boot, there is typically no fat cat getting rich on the backs of hard-working farmers behind the scenes; most farmers markets are nonprofits run by volunteers or part-time staff for nominal pay. Some of the beneficial effects of farmers markets on communities include:

- Stimulating the local economy by serving as an incubator for small businesses and by increasing activity in the commercial areas where markets are held
- Creating conscious consumers and raising awareness of the importance of sustainable agriculture by exposing the public to small farms
- Increasing consumption of fresh produce across the board
- Increasing access to fresh food for those in need, with many now accepting SNAP benefits (also known as food stamps)

These are all *very good things*.

To broaden their appeal, many markets have added local arts and crafts vendors, a topic of some controversy among farmers who believe their presence detracts from the market's purported mission to be a forum for the community to buy locally produced food. In our experience, most "farmers" markets today work far better for those types of artisans than for farmers. I humbly suggest that the term "local market" is a better fit for these venues. While I will happily shop at these markets for the rest of my life, I would never sell produce at one again.

Some farmers sell at the market because they simply choose to do so and always will, but many don't know any other way to reach customers. If you're a market gardener with a good spot at a well attended market where you're able to sell everything you grow and you enjoy it, you're probably not reading this book. But it should not be automatically assumed that to grow and sell produce directly to consumers, you must go to the farmers market, as is still the default for so many. For Pure Land, that model was unsustainable. We want to show those who find themselves in a similar position that there's a way to make a living growing food without the farmers market.

Here are the reasons farmers markets didn't work for us.

Ripe melons wait for no man. As experienced farmers know, many crops picked ripe, at their peak of flavor, do not have the shelf life of the practically indestructible produce found in a grocery store. Fruits and vegetables destined for distributors are often picked well before ripening, rendering them lacking in flavor and nutrition but very shippable and long lasting, features considered much more desirable to a distributor. In fact, much of the effort in developing new improved varieties has been on improving characteristics such as yield, durability through shipping, and shelf life, as opposed to taste and nutrition.

The disparity in flavor between what you'll find at the grocery store versus your farmers market is best illustrated with cantaloupes, or more broadly, muskmelons. A muskmelon picked ripe is indescribably superior to what most people are used to eating. Their flavor can be exceptionally sweet and complex, and they are exceedingly fragrant. Ripe muskmelons can be smelled for miles. One day, I was making some restaurant deliveries that included a tote of melons. At one stop, the chef was about a hundred feet away when he caught a whiff from inside my truck and exclaimed "Wow! You got melons in there?"

When ripe, a muskmelon will crack around its stem and release from the vine with the gentlest tug, called "slipping". There is an unexplainable phenomenon whereby the majority of a muskmelon crop will always slip on Sundays, the day after the farmers market, leaving the market gardener stuck trying to hustle them midweek or store them for the following weekend while they stink up the place and go soft. Even if kept at the optimal storage temperature, the quality of a melon picked ripe diminishes in just a few days and they quickly become unsellable. Some

Harvest a muskmelon at full slip for best flavor.

crops are just not designed to sit around for a week. If you consider the flavor to be the most important aspect, as we do, picking them unripe to increase the storage life is not an option.

Some try to get around this issue by going to a midweek farmers market, but it's not just the time standing at the market table that is sacrificed. It also takes time to pick, prep, store, and load everything, set up and break down the tent and tables, and finally, find buyers for what did not sell because midweek markets are usually poorly attended. All this easily adds up to more than a day spent not farming.

We should care more about the quality and flavor of what we grow and eat, and the way to inspire people is by appealing to their senses. Pure Land customers talk about our cantaloupes year-round on our social media because they'd never tasted one that was picked ripe before finding us and had no idea how good a cantaloupe could be. We're not going to convince people to support local food if they feel like they can get the same thing cheaper and more conveniently at the grocery store - we have to show how much better locally grown food can be, and that starts with serving it at the peak of flavor and quality.

A four-hour window on Saturday mornings is prohibitively inconvenient for a lot of people who would otherwise buy from you. Farmers markets are generally held for three or four hours once a week, creating extremely limited availability of the products to the customers. A lot of people want to support farmers, but Saturday mornings are a busy family time spent making French toast and shuffling to soccer games. It is just too inconvenient for many to patronize a business that is only open once a week for four hours on Saturday mornings.

We know farmers that have abandoned markets in favor of running their own full-time retail, but that presents plenty of other problems. If you decide to go it on your own, your regular customers must be willing to go out of their way to buy from you. Many may first look for you at the market but settle for purchasing from another farm out of convenience, since they're already at the market anyway. The structure of business hours at a farmers market at least creates a schedule that customers are willing to follow; when farms decide to handle their own retail, some customers tend to believe they're always open. Harvesting all through the week and managing inventory are time consuming and disruptive to a farming schedule, while setting limitations on availability risks losing customers. Farmers who offer orders for pick-up regularly lament customers not showing up, despite being given several days and times to choose from. Then they're faced with restocking and managing refunds. Have you ever successfully restocked a tomato?

Weather is utterly uncontrollable but dominates market performance. This is frustratingly basic, but most farmers markets still take place outdoors. If it rains during the market, the customers don't skip eating that week. They head to the grocery store. *You* are the one soaking wet with a week's worth of perishable product and an empty cash box. Frankly, I'm on the customer's side here; when it's raining, very hot, or very cold, it's unreasonable to expect people to shop outside. I certainly never enjoyed spending five hours outside in 30 degree weather to sell six heads of lettuce, and I don't blame anyone for not joining me. Our last year of markets, it rained every Saturday morning in June, the peak season for pro-

Rain on Saturday morning is a real bummer.

duce in North Texas. I'm still not over it.

There are too many markets. Every neighborhood wants their own market. Farmers markets are social events where connections are made not just between producers and consumers, but neighbors. You'll probably never text a friend to meet you at the grocery store to do your shopping together, but a trip to the farmers market entails strolling outside with your dogs, a steamy locally roasted coffee in one hand and breakfast burrito in the other. There's probably a pony ride or face painting for the kids and you might pick up some vegetables while you're there, too. *It's fun*, I get it. But from the seller's perspective, creating more markets dilutes the number of shoppers at each one, leaving vendors saddled with purchasing additional market setups of tents and tables and paying someone to take product to other markets. One rancher I spoke with said he had to send staff to four markets to make what he used to earn at one market ten years ago, leading him to pull out of them entirely and move to wholesale. We don't need more markets; we need better ones.

Should you decide to set up and staff booths at several local farmers markets, don't necessarily expect the same sales as the one you, the farmer, work. Shoppers find it easy to tell the hired help from the farmer, and they want to talk to the farmer. Staff that can't describe the farm's policy on pest control or give gardening advice can be a turn-off to customers.

They come with built-in direct competition for your products. Shoppers rarely make significant purchases from any single vendor at the market because they're faced with so much choice. A typical farmers market customer might buy carrots from me every week, arugula from John, and strawberries from Pam, even though all three of us have mostly the same produce on our tables. As one of my regulars told me, "I want to support all my farmer friends!" It's hard to argue with that.

More problematic are farmers markets that permit the resale of produce purchased wholesale from distributors, often grown in California, Florida, or Mexico, allowing vendors to dupe customers into believing it's locally grown. When revealed, this practice is usually as shocking to market customers as common, which is to say, very. It is an immense disservice to the local food movement for a vendor to display red tomatoes and watermelon at the first market of spring when the local growers have only kale and cabbage. If market management will not commit to being grower-only or enforcing the market's procurement rules, the vendors often turn to policing each other, creating a toxic environment of people desperate for sales and a pretty miserable place to work.

It's hard to tell your story. A bunch of carrots looks like a bunch of carrots, whether it's on a table at the farmers market or in the cold case at the grocery store. What makes your carrots different is their story. Maybe it's an heirloom variety you're trying for the first time, and you had trouble getting them to germinate in your crusty soil, so you had to baby them and hand water for a few weeks, but they turned out to be the sweetest you've ever grown. Your granddaughter can't stop eating them raw, but your favorite way to fix 'em is roasted in a hot oven with a drizzle of maple syrup.

It's hard to get a personal story like this across in a ten-by-ten tent in a parking lot, a place that feels about as far away from a farm as you can get. This is why you see many savvy farmers hang large banners with pictures of their farm in their *booths, to visually make the connection in customers' brains that I, the person standing in front of you and* pictured here on a farm, grew this food on the table with these hands at my farm. Absent the story, your carrots look just like the ones at the grocery store and the higher price tag doesn't seem justified. Ask any grower at the farmers market, they all have stories about these shoppers. "Wow, that's so expensive, I can get those two for a buck at the megamart!" Why would the customer pay more for what they perceive as the same thing they can get cheaper elsewhere?

The success of the market is largely out of your hands. A well-managed farmers market secures a good location, has engaging social media, maintains an informative and updated website, and makes the effort to promote their vendors as much as themselves. They schedule special events and musical acts over the season to keep it fun. They get involved with the surrounding community to advertise the market and draw people in every week. Good markets make deals with nearby businesses to provide ample parking and curate a balanced mix of vendors to ensure there is good variety. Bad markets often do none of these, limiting their general appeal to shoppers. Of course, there are tactics you can use to boost your sales. A colorful, bountiful display never hurts. The vendor should always stand, never sit, and greet every shopper as they approach the table. Be ready to take any form of payment so the shopper can spend freely without having to conserve their cash. But ultimately, these tips will only get you so far. If the market stinks, people won't want to come.

Even inside a relatively well-run and busy market, there can be issues. It isn't possible for every vendor to be front and center, so there may be farmers with prime booth spots that have great sales while others are tucked in a dead end no one ventures down. Markets that prioritize supporting local agriculture reserve good spots for local farmers, but these are few and far between. The vendor mix can be an issue, too. It's obviously unwise to join a relatively small market where there are already plenty of farmers. Unbalanced the other way, a market with twelve craft vendors and only one farmer might seem attractive but probably won't bring in many shoppers looking for produce.

Farmers markets require consistency to be successful, both for the individual vendor and for the market at large. It is in the market's interest to have the same roster of farmers there every week for as long as they can because it offers consistency and legitimacy to the customers. Market managers justifiably hate having to tell shoppers that the farmer they're looking for isn't there that week, regardless of the reason why. Many markets ask that you sign a season-long contract for discounted rates or pay a fee for dropping out early. New markets especially will make a hard play for the vendors' commitment, and if it's in a convenient location or you have reason to believe strongly in the potential of the market, it may be really tempting to sign on.

On the vendor side, it takes time to build up a base of regular customers every week, and regulars are every vendor's bread and butter. Our most profitable and in-demand crops are the ones harvested in the spring and early summer, which is when farmers market attendance peaks here in North Texas. If we had to be present at every market of the season, we'd have to sacrifice some of the spring and summer crop growing space for less lucrative fall and winter crops, effectively putting a damper on our earning potential for the year.

Except for at the very best markets, there are just not enough shoppers. This is in part because there are too many markets, but markets are also facing downward pressure on attendance from new forms of competition. Prepared meal delivery services and kits are gaining in popularity, and some even boast of buying local produce in their marketing, regardless of the consumer's location (obviously some kind of sorcery). Many of the folks who still cook from scratch choose the relatively new convenience of ordering groceries online and having them delivered. All of these incredibly easy, efficient, and exciting new ways to buy groceries result in fewer and fewer people heading out on Saturday mornings to buy local.

The Limitations of Other Sales Outlets

Other common avenues of sale for a market gardener include wholesaling and operating a CSA. While there are positive aspects of both, they remain far less efficient and lucrative when compared to the pick-your-own model.

Wholesale. Wholesaling is when a farmer sells their produce to a third party, who then sells it to someone else. Most small farmers tend to focus on one sales model or the other: wholesaling to restaurants and distributors, or retailing at the farmers market. Unless a farm is of considerable size, it's hard to do both well. Wholesale prices are usually much lower than retail, meaning it's necessary to produce larger quantities in a streamlined manner to match the revenue from a smaller volume sold at retail prices. Like many growers of our very small size, we considered these outlets complementary; we would sell wholesale to distributors or restaurants at a reasonable discount when we had a surplus but prioritize farmers markets to capitalize on retail prices. After all, it took the same amount of effort to grow and harvest regardless of where the produce was sold.

Thy small farm shalt:

SHUN MIDDLEMEN AND MIDDLEWOMEN LIKE THE PLAGUE— FOR THEY ARE A CURSE UPON THEE.

Not all wholesale opportunities are created equal. Produce distributors may offer some stability in terms of consistent sales, but restaurants operate on razor thin profit margins, often closing with little to no notice. This prompted us to focus on building relationships with chefs, rather than restaurants. The contact profiles of some of the chefs in my phone reference restaurants three and four closures ago. True farm-to-table chefs are fun to work with because they get excited by all the unusual stuff you can grow that they can't find anywhere else. Embrace this but be wary of any time a chef tells you they'll take "all of" any particular crop you can grow; we came to learn "all of" often means far less than you're actually capable of growing. You just might find yourself stuck pushing a ton of weird white beets at the farmers market that most shoppers aren't willing to take a chance on.

Wholesaling holds some appeal because the demand is more consistent and it can move larger volumes of product at a time, but handling the distribution should be a major consideration. Once you have a roster of restaurants, you may find yourself running around town a whole day or more per week making deliveries, keeping you away from farming. A pair of local farmers in our area teamed up to solve this issue, with one taking the others' restaurant deliveries on Mondays, and the other taking both of their deliveries on Thursdays. This grew into a food hub that now distributes products from over sixty farmers to dozens of restaurants, allowing participating farmers to make one drop off at the hub and head back to their farm to work. It's a major time saver.

CSAs. Community Supported Agriculture (CSA) is a system where the farmer projects their output and offers a number of "shares" for sale at the beginning of the season. "Members" who buy in receive a portion of the harvest periodically over the growing season, usually weekly or biweekly.

Although the ideology has drifted, the intention of a true collaborative CSA is for members to invest financially in and get to know the person who grows their food, which is a wonderful concept. Back in the day, many CSAs required members to help with the farm work or management but today, this aspect has been largely abandoned as too many members are unwilling or unable to satisfy such commitments. Most of the farms now offering a "CSA" simply use it as marketing, a way to sell a higher volume of produce in one transaction. This helps to offset the increased labor spent to prepare the shares.

An attractive aspect of CSA growing to the farmer is the infusion of cash from sign-ups at the beginning of the season, which can pay for growing supplies and equipment. If you can handle the stress of taking people's money and facing a season's worth of high expectations, you have a stronger stomach than I do. Unfortunately farms that struggle to finish a season with little to no profit don't have enough money left to prepare for next season. This system allows them to keep going.

Anecdotally, many CSA farmers we know that had wait lists for membership in the aughts can no longer fill their shares. In our globalized world, consumers expect variety, regardless of the season and location. Many are just not satisfied with the several months of chard and cilantro that come with genuine, seasonal CSA memberships. The explosion of delivery services marketed as "farm boxes" may be the last straw, where products are aggregated from a number of farms in a central location. Consumers select the items they want, which are then packaged and delivered by a third party. Individual farms cannot compete with this. For a CSA farmer, creating custom boxes or filling special requests drastically complicates fulfillment, and spending one or more days every week making deliveries is out of the question. A few years back, I attended a workshop with a CSA farmer in Austin who implemented a strict no-replacements policy. She confided that she had started to tell her members, "Every replacement you request is five minutes I can't spend with my children. We have 300 members. You do the math."

We started our farm with the intention of becoming a CSA, but quickly ran into the challenge of providing sufficient variety. To keep members happy, you must be able to deliver a wide variety of produce in practical quantities. If you have rich

A typical CSA share.

loamy soil and California temperatures, you have no excuse. (Kidding! Mostly.) But it can be impractical to the point of cost prohibitive for a small farm to try to create the conditions in which they can grow "everything." Some farmers succumb to growing loss leaders or crops with marginal production just to bolster variety, which comes at the added cost of space that would have been better used for a more profitable crop. This also leads many small farms to buy produce from each other to bulk up their boxes, which is good in the sense that collaboration tends to have a net positive effect eventually (like our two farmers taking turns on deliveries) but bad in the sense that they now have business partners whose actions are out of their control. If you're counting on buying the extra pears from the farm down the road for your shares but they had a great market and sold out, you're in trouble.

The Agritourism Solution

Now that we've assessed the limitations of traditional direct-to-consumer sales avenues for the market gardener, let's talk about a solution.

At its core, agritourism is the combination of agriculture and tourism. The farmer provides an appealing educational or entertainment activity on the farm, and the consumer pays a fee to participate. According to a 2018 article in the *Journal of Agriculture, Food Systems, and Community Development*, agritourism is defined as activities falling into one or more of the following categories:

- direct-to-consumer sales (e.g. farm stands, pick-your-own)
- agricultural education (e.g. school visits to a farm)
- hospitality (e.g. overnight farm stays)
- recreation (e.g. hunting, horseback riding)
- entertainment (e.g. hayrides, harvest dinners)

In this book we will focus on what we know best, a pick-your-own fruit and vegetable farm (also commonly stylized as you-pick, u-pick, or u-pick-it). Before we go any further, I want to distinguish between agritourism and agritainment. They are often used interchangeably, but it's generally accepted that agritourism involves an educational component, while agritainment focuses primarily on recreational activities that just so happen to take place on a farm. Farm tours and pick-your-own are considered agritourism, while corn mazes and fall festivals are agritainment. At Pure Land, we don't do agritainment. *We grow food.* Our type of agritourism is for farmers who primarily want to be *farmers*, although there absolutely are some teacher, tour guide, event venue director, and clown elements rolled in.

In the United States, the first documented agritourism ventures were dude ranches in the 1880s. These operations offered city slickers a taste of the Wild West by allowing them to participate in some of the more romantic duties of farm work, like rounding up cattle on horseback and camping. In the 1920s, the number of people living in cities surpassed those living in rural areas for the first time. Automobiles had recently become much more accessible, and so "visiting the country" was born as a new weekend pastime. Desperation to escape the daily stresses of the

Great Depression and World War II boosted enthusiasm for these activities into the 1930s and 1940s. The surge of health consciousness in the 1970s that gave rise to a boom in farmers markets swept agritourism into broader popularity as well. It continues to grow; the USDA's Census of Agriculture reported that farm revenue from agritourism more than tripled from 2002 to 2017.

Like farmers markets, agritourism can have a number of beneficial effects:

- Allowing farms to increase and diversify their income streams. Farms that only produce for a limited season can extend the farm's period of income generation with off-season agritourism opportunities, like farm stays or dinners on the property
- Increasing profitability which creates opportunities for a farm to consider expanding and hiring more employees
- Bringing the positive economic impact of tourism to rural communities, where it can be especially impactful as tourists often patronize additional locally owned businesses during their visit
- Creating a positive feedback loop of desire to support local producers. A 2020 study of agritourism participants at six farms in North Carolina found that participants reported increased intentions to buy more locally produced food after their experiences [Brune 2020]

Agritourism allows farmers to educate the public about conservation and regenerative agriculture through the power of fun, immersive experiences. Let's face it - to most people, agricultural and environmental policy are pretty boring. Some folks will have heard of the Farm Bill, but they probably don't know what's in it or how it might actually affect small local farmers in practice. I don't say this to admonish them; the majority of people in society do not personally know *a single farmer*, especially in urban areas. They only see farms in children's books, primary-colored pages of old men in overalls with an oink oink here and an oink oink there. But when a real farm becomes part of a community, they organically begin to pay more attention. Our work shirts say "Farmer Jack" and "Farmer Meg" for this very reason. When a personal connection is made to a real life farmer and farms become places they take their families every summer, they care more about the things that affect us farmers, which builds advocacy for policies that support local, regenerative agriculture.

It's important to recognize that the tangible goods sold through agritourism, *food*, are not the same as similar goods sold elsewhere. Industrial agriculture can grow USDA Certified Organic produce anywhere at any time of year, and they do it at tremendous scale. They can deliver it right to the customer's door, sometimes in hours. They can always beat small farms on price, and they have infinitely more capital at their disposal. But there is one thing they cannot compete with us on: the experience we can provide on our farms. Agritourism is immune to competition from large-scale industrial agriculture because the product sold is fundamentally different. We aren't just selling carrots, and we certainly aren't selling convenience;

we're selling a morning on a farm with a couple of friendly farmers and access to the freshest produce money can buy.

Converting our farm from a market garden into an agritourism venture has had an overwhelmingly positive effect on both our business and our lives.

Agritourism solved every one of the problems we faced at the farmers market.

- Pick-your-own farming is incredibly efficient. Produce is sold the moment it's picked, so the customer gets to enjoy it at the absolute peak of freshness and it does not require storage or transportation by the farm.
- The open hours are determined by the volume of produce ready to pick, which works out to four or five mornings a week during our picking season. This gives ample opportunity for most people to find a time that works for them to visit. If there's plenty of produce ready to pick in the field, it's easy to bring in more people by opening more spots, hours, or days on the fly.
- While we have not yet figured out how to control the weather, if a pick is rained out, it can be rescheduled for whenever the storm passes or the next day. It is not necessary to wait a whole week until the next opportunity for retail sales.
- Competition is low because there are very few other pick-your-owns in our area (although I don't believe it would matter much if there were, considering the demand) and obviously, no produce from another farm to choose over ours at the pick-your-own. We are no longer in competition with grocery stores and commercial-scale industrial agriculture because what we sell is an experience they do not provide.
- Our success lies wholly within our hands and is not dependent on any larger organization to attract customers or promote us. We operate independently, enabling us to make the right decisions for our business without having to consider impacts to a partner or parent organization.
- We can set our open season around when the most profitable crops are ready.
- We don't have to spend time away from farming to prepare orders, go to markets, or deliver to wholesale customers.
- Perhaps most consequential, there are more people who want the experience and wish to buy from us than we can supply.

Sharing the physical work of farming is good for your body. Growing, picking, prepping, and bringing produce to market is physically exhausting. To the market gardeners reading this, how many late nights have you spent hunched over after a backbreaking day of work, furiously bagging greens for market after your family has gone to bed? Or, after loading the truck, driving an hour, then unloading and thoughtfully arranging hundreds of pounds of tables, tents, and produce at the market in the wee dark hours before dawn, how many times have you lost a sale when a customer asked you if your wares were picked that morning and you snort laughed coffee through your nose? (Only once.) We're moving as fast as we can.

There is still plenty of manual labor to support a pick-your-own farm, but delegating the harvesting and marketing tasks to the customer, frankly, is an immense physical relief. As of the 2017 USDA Census of Agriculture, the average age of an American farmer is 57.5. Fifty may be the new thirty, but there are limits to what we should put our bodies through to make them last the rest of our lives. We grow some 40,000 pounds of food every summer; if not for the pick-your-own model, we might pick up and put down those 40,000 pounds ten times before they were sold.

A picker loaded down with baskets of food

It takes less time overall. The labor it takes to bring produce to market is not only physically challenging but extremely time consuming. At the scale we were growing when we used to sell at farmers markets, which was smaller than our current scale, it would take one full-time employee about two days to pick, process, and market one-to-two thousand dollars' worth of diversified produce. Now that we're pick-your-own, we've reclaimed two whole days every week to spend growing more food.

Not every task will take less time in the pick-your-own model, but the total number of hours never come close to the time needed to harvest, prep, and market the old way. Answering emails and messages, updating the website, and posting to social media are all daily chores that we spend more time on than we used to, an inevitable consequence of managing an experience for so many people. The distinction is that unlike most market gardening tasks, these can be performed from a comfy chair with a cold beverage.

The pick-your-own model saves time on a daily basis but can also have a more sweeping effect if you choose. We'll get more into this later, but we have reduced our growing season from year-round to about five months of full-time farming for the two of us, freeing up the rest of the year to pursue other interests.

It is incredibly fulfilling. There is no better feeling than being thanked for your hard work. It used to happen at the farmers market on occasion, but it happens every single day at our pick-your-own. Moms who arrived with not a hair out of place just an hour earlier often reach the checkout tent soaked in sweat, wrangling a squealing crew of muddy children - some now missing their shoes - and say with absolute sincerity, "*Thank you so much* for letting us come to your farm! We had *the best time*." Our pleasure, ma'am.

Not to get too touchy-feely, but something about the experience of being out in nature, sweating, harvesting fruits and vegetables seems to arouse our humanity. Everyone is a little kinder to each other at the farm, more patient, and helpful. We're often treated with undue awe by our visitors, especially those who have tried their hands at gardening. But when you think about it, it *is* pretty marvelous to grow real food from tiny seeds to feed people. There is no one immune to the wonder of it, and it makes for a very rewarding way to spend your time.

Younger generations and the planet benefit. It's one thing to read an article about the value of regenerative agriculture, but quite another to experience it in a visceral, personal way. We spend a lot of time educating our guests on the importance of small farms like ours that use sustainable practices because every visitor is a potential new conscientious consumer, someone who might think twice about where the cheap meat from the grocery store came from and look for a local producer instead. We need way more of those people. The folks who choose to visit a pick-your-own farm are going far out of their way to opt into listening to what a farmer has to say. So, talk!

The magic of discovery!

Children in particular benefit from exposure to gardens and farms, and most of our guests are families with kids. For many of the parents, let alone the kids, visiting us is the first time they've ever been to a farm. Children who experience hands-on learning in school gardens eat up to three times more fruits and vegetables than those without. We see it every day at the farm; a kid in the checkout line firmly gripping the carrot he worked so hard to pick, his *Excalibur*, saying "Can I eat it yet? Can I eat it?" And as soon as it has been weighed and washed, it's down the hatch. Toddlers waddle up from the blackberry field with apologetic parents, their happy little faces stained purple from the eyes to the chin. We are planting seeds in their little brains, initiating the joy of healthy eating and respect for nature and the planet.

We make more money. Critically, *it is profitable*. The bottom-line effect of pick-your-own farming is not merely being able to sell the same amount with less work than going to the farmers market; it's the ability to sell significantly *more*. Now that we can focus most of our efforts growing more food, our revenue has increased to a level that supports the two of us without having to work harder or get bigger. It's a beautiful thing.

Chapter 2

UNDERSTANDING AGRITOURISTS

As the self-appointed ambassador of pick-your-own farming, I talk to a lot of farmers about getting into agritourism. Most are intrigued by the prospect of decreasing their workloads and increasing sales, but the part they always get hung up on is opening the gate to let customers into their farms. *Who are these people? What are they gonna do to my place? And why on earth do they think this is fun?* Fair or not, farmers have earned a bit of a reputation as misanthropes, but their aversion to letting the sometimes litigious general public onto their property is understandable. If you are going to invite someone into your farm, which is often also your home, it's reasonable to want to know a little more about who they are and why they're there. The stakes are clearly higher than if they were merely browsing your table at the farmers market on Saturday morning and moving on.

We've spent a few years observing the people that choose to visit us; aka, agritourists. Some major trends have emerged we believe you're likely to encounter anywhere, but every market will have idiosyncrasies due to demographics, proximity to urban areas, etc. Overwhelmingly, we've found that agritourists are earnest people with sincere motivations to visit your farm that are likely aligned with your own motivations for being a farmer. There is little to be afraid of; agritourists are not motivated by destroying plants or mauling themselves on a tractor. After some 25,000 guests over the years, I can count on one hand the number of truly terrible experiences we've endured from another human being, far fewer than I've encountered in the rest of my daily life.

Not everyone will be interested in participating in agritourism, and that's okay. When asked by a fellow farmer what we raise in Collin County, I once heard a friend of mine reply, "Houses." We live in one of the fastest growing counties in the United States, and urban sprawl has engulfed our farm over the last few years. The quiet dirt roads around us were paved a year or two after we arrived, then widened.

Naturally, it's a bummer to see a suburb where a forest used to be, but our attitude is that every new house is a potential new customer. *Everybody eats.* But not everybody is willing (or able) to come pick the carrot themselves. Heck, not everybody is even willing to cook the carrot themselves these days, as seen by the recent rise in demand for prepared, ready-to-eat meals. The good news is, we don't need *everybody.* At our scale, growing three or four acres of diversified fruits and vegetables, it only takes 5,000-10,000 visitors over the season to pick everything we can grow, a drop in the bucket of the nearly 8 million residents of the Dallas-Fort Worth Metroplex. At most, just over a tenth of one percent of them. Piece of cake.

Of course, the more people we reach, the more we can expand public support for regenerative agriculture and local food systems, which is the overarching goal. It's unrealistic to expect consumers to change all their buying habits at once, but every decision to buy local is a step in the right direction. On the individual level, we should foster easy, incremental changes that encourage the public to support the local food system more often, and our job as local food producers is to present the public with appealing opportunities to do so. I'm reminded of a quote by Anne-Marie Bonneau, the Zero Waste Chef, regarding a social media post by a woman who had minimized creating trash to such a degree that her entire output for the year fit into a one-pint mason jar. Admirable, but unattainable for most of us. In response, Bonneau said, "We don't need a handful of people doing zero waste perfectly; we need millions of people doing it imperfectly." The same is true for creating a robust local food production system; we don't need *a few* people getting *everything* they consume from local farmers and ranchers, we need *millions* of people buying from them at least *some of the time.* This isn't to say I don't appreciate our veteran regulars and true believers; they are vitally important to our business. There just aren't enough of them yet. The first step to cultivating more buyers is to understand who these potential customers are and what they truly want, which we will explore in this chapter.

Agritourism hasn't had the marketing push into cultural ubiquity that farmers markets have, but I believe we can change that. Most people still think of agritourism as pumpkin patches and apple picking, seasonal family activities that are attended annually and promptly forgotten about until next year, save for the photo on the fridge. With more farms like ours offering the opportunity to participate in supplying consumers with real groceries - not thirty pounds of apples at one time, but practical volumes of commonly consumed fruits and vegetables over a season - I believe we can propel pick-your-own farms into the mainstream. The more farms there are to visit, the more this kind of agritourism will enter our greater cultural awareness and the larger the customer base making regular purchases from local farms will grow.

Agritourists' Motivations and Values

A Difference in Perceived Value

A friend of mine once owned a small organic grocery store in Dallas. She described watching shoppers at the farmers market "acting like they were using Monopoly money" compared to the frugal behavior she observed in her store. The produce she stocked was purchased from local growers at a wholesale rate, then marked up to about the same price it would be on the farmer's table Saturday mornings. I told her I could easily fetch a dollar apiece for slender spring leeks at the market, a price she thought would be impossible in the store.

In marketing lingo, perceived value is a customer's assessment of a product's worth or desirability. This concept is carefully considered by marketing professionals in circumstances where customers have a lot of choice; the product's marketing will try to convince the buyer to spend more for this particular brand of paper towels because they're *worth* more. The marketing might present studies that show they're more absorbent, or declare they were produced ethically and without synthetic dyes, or maybe just give it a snazzy-looking label and let the superior aesthetics reel you in.

The perceived value of my leeks dropped like a rock when they hit the cooler shelf in my friend's grocery store. I, the farmer, was nowhere in sight, and despite the store marketing the produce as locally grown, the customer couldn't connect the leeks to my farm to find them worthy of the farmers market price. Even though the customer had chosen to go to an organic grocery and surely expected their bill would be higher than at a conventional supermarket, it's *still* a grocery store, and as we'll see, shoppers in a grocery store are not motivated by supporting local agriculture.

Agritourists take this Monopoly money mentality to the next level. Making a sale at the farmers market is not particularly difficult - the customers have chosen to be there because they're looking for exactly what's being sold. It doesn't take a marketing genius to sell vegetables to people shopping for vegetables. But agritourists are not only purchasing locally grown produce, even if that's the only tangible thing they walk away with. The perceived value of what they're getting is greater; they're buying access to the farm and farmer, quality time with their family, fresh air, and locally grown produce. They're comfortable spending more because they acknowledge that they're *buying* more.

Another boost to the perceived value of our produce is that the context in which it is purchased - through participating in the harvest on the farm - changes the customer's perception of how much money is "reasonable" to pay for it. Years before we transitioned to pick-your-own, one of my aunts came out to help us harvest one day. When we finished, she confessed that she used to think our prices were really high, but after spending *one afternoon* participating in the farm work, she thought we should raise them! Requiring that customers do some of the work involved in growing food makes the food more valuable. I bet we'd all be willing to pay a little more for our t-shirts if we spent a few hours working in the factory.

Anecdotally, we are haggled with far less often at the you-pick than we were at the farmers market. Every Saturday morning would turn out some bargain shoppers, real wheeler-dealers proposing that you sell them two $3 bunches for $5 or making a lowball offer on your remaining produce near the end of market. Not only is that kind of bargaining extremely rare at the you-pick, but people are spending far more money, an average of ten times more than our average transaction at the farmers market. They're spending more *and* haggling less, which feels counterintuitive unless you consider the difference in perceived values.

Surveyed Motivations and Values

People who simply want to buy food have no shortage of options. They can go to supermarkets, wholesale clubs, specialty grocery stores, co-ops, farmers markets, even place an order online with a few taps on their phone. What motivates a shopper to choose one outlet over another?

Let's start by comparing grocery store shoppers with those at the farmers market. Among other factors, consumers who choose the grocery store are primarily motivated by low prices and convenience. Grocery stores operate at economies of scale, which allows them to offer goods at lower prices than most small businesses can. There are also numerous grocery outlets to choose from in most areas and they're open nearly round the clock, so you can get broccoli, dish soap, and cat toys all in one stop at 10pm if that's convenient for you.

Shoppers who prefer the farmers market are typically not motivated by cost and convenience. One study of 1,488 farmers market attendees found that 78 percent of them cited purchasing fresh produce as their primary motivation for going [Gumirakiza 2014], a statistic backed up by numerous other studies. Other reported motivations of farmers market shoppers include supporting local agriculture, seeking "safe" food not produced by the industrial food system, minimizing their environmental impact by purchasing food produced in an environmentally responsible way, and desire for social interactions both with vendors and their fellow shoppers.

That said, farmers market shoppers are not our target customers anymore. How do they and grocery store shoppers compare to agritourists? While the habits

Grocery store shoppers are motivated by cost and convenience, while agritourists are motivated by finding the freshest produce and supporting local agriculture.

of farmers market shoppers are fairly well understood, there is less data available on what makes agritourists tick. Still, we can make some qualitative observations and inferences based on the existing data and different natures of these ways to buy food. If a farmers market shopper is motivated above all else by desire to find the freshest produce and support local agriculture, it seems reasonable that an agritourist going directly to the farm to pick is *even* more strongly motivated by those potential outcomes. Buying food through agritourism may also be more expensive as it includes a premium for the tourism aspect, and it's certainly less convenient than stopping at the market in their neighborhood where there's a variety of goods for sale. Therefore, one could infer that agritourists are even less motivated by low prices and convenience than farmers market shoppers.

I wondered how agritourism participants would rank their motivations for visiting *our* farm, so I conducted an online survey of our customers. The question posed was, what are the most important things you're looking for when you come to Pure Land Farm? I listed the motivations cited by farmers market shoppers as answer choices, adding ones I suspected might be applicable to agritourists - a fun activity and spending recreational time outside in nature - and allowed them to select all that applied.

The number one motivating factor for agritourists in choosing to visit our farm was to support local agriculture and build a personal relationship with a farmer, followed by finding the freshest possible produce. A motivating factor I didn't anticipate or include in the survey topped the write-in responses: the desire to educate

Reported motivations of Pure Land customers in 2022.

their kids on where food comes from.

Spending time outside in nature, consuming "organic" or "safe" food, and supporting food produced in an environmentally responsible way are desired outcomes that came in lower, but were still reported by more than half of our guests. Our customers often express gratitude for our plainly evident regenerative growing methods, both because they value food safe from the perceived dangers of industrial agriculture and because they appreciate the positive effect these practices have on the environment. This data and our anecdotal experiences support the notion that agritourists highly value both the environmental and individual health impacts of food produced regeneratively. Therefore, the use of regenerative growing techniques just creates a more desirable product. Customers often ask what kind of growing methods our farm uses, and I've never disappointed anyone with my honest answers.

Unsurprisingly, only 17 percent of respondents reported being motivated to visit our farm by low prices, and fewer than 8 percent by convenience. Farmers market shoppers and agritourists know that buying from local farms costs more than the grocery store and choose to shop there anyway. A 2005 study out of UC Davis corroborated this by finding that two thirds of respondents who purchased products at a farm during an agritourism activity were willing to pay equal to or more than they would for the same items in a supermarket [Jolly 2005]. The same study found that 68 percent of respondents were willing to pay between $1 and $15 in entrance fees to access the farm and activity. We'll talk more about this in Chapter 6, Running your Picking Season, but I mention it here to plant the seed that there is no need to price products too low or waive entrance fees out of fear that people won't come.

Now that we have established agritourists as being motivated by supporting local agriculture and forming a personal relationship with a farmer, finding fresh high-quality produce, spending time outside in nature, and doing a fun activity, what are the values or criteria they use to measure whether they're reaching these desired outcomes? Of these, which are valued by the most people? Understanding what the customers value tells us where to focus our efforts to appeal to the broadest group of people. Failure to satisfy one or more of these values can make it less likely that agritourists have a good time and ultimately, they may decide not to come back. The survey question we asked the same group was, what do you like about Pure Land Farm? The suggested answer choices were gleaned from several studies reporting agritourists' values and our assumptions.

The organic, high-quality nature of our produce appears to be the number one measure by which our customers decide if they've successfully reached their desired outcome. It's an important reminder that a farm can have everything else going for it, like being in a prime spot with lots of parking and perfect weather, but if the lettuce is full of holes, the customer is unlikely to perceive higher value in it. Across the board, shoppers expect high-quality produce when supporting local agriculture; *nothing* on a pick-your-own farmer's to-do list should come before taking care of

Values

Reported values of Pure Land customers in 2022.

the plants. Our focus has naturally fallen on the top two values - keep the produce quality high, and make it fun. But we also spend a lot of effort on keeping the farm beautiful and allowing people to get to know us, too. Again, the leading write-in response to this question was an appreciation of the way we help their children learn about growing food. Clearly, agritourists wish to raise their children to support local agriculture as they do, which is heartening and critical to small farms persisting into the future.

It is notable that "getting to know my farmer" was not the top reported value, considering that sentiment was part of the top motivation. That suggests to me that the lived experience of a "personal relationship with a farmer" doesn't describe a straightforward friendship, but establishing a bond that encompasses what a local farmer represents. If asked what comes to mind when one thinks about "local farms," a person might say fresh, healthy food, wide open spaces, caring for the environment, small family businesses. Conceivably, our customers are not really motivated to *hang out* with us (although we are awesome), but to patronize a business they perceive as honest and fair, one they can trust to sell them high-quality food produced in a responsible way and provide an enjoyable, safe environment for their families to visit. We are essentially inviting customers into our home, which puts them in a vulnerable position. I, myself, am certainly more comfortable visiting the house of a friend than the house of a stranger, eating food my friend has prepared and letting my children run around my friend's big backyard. A "personal relationship with a farmer" can provide this comfort.

We have come to believe the fundamental driver is that agritourists want to belong to a community that includes the people who grow their food and the places they do it. A favorite customer once caught a clip of me on the news and said she jumped off the couch, pointed at the screen, and yelled, "Hey! That's my farmer!" *My farmer*. I love this story not only because it's so funny to picture her doing this, but it really reveals the depth of the bonds we've created through our farm. The word "my" is reserved for our most intimate relationships; my wife, my kid, my best friend. The "my" implies trust, commitment. In this line of work, I feel like "my farmer" is the highest honorific we can hope to achieve.

In Their Words

A limitation of the survey approach to measuring our customers' motivations and values is that we've suggested potential answers to the questions. To see what our guests have to say without putting words in their mouths, we created a Word Cloud from our online reviews. Word Clouds reveal the most frequently used words from a body of text, sizing them in proportion to the frequency with which they appear. After filtering out the common insignificant words from our last three years of reviews (about 120 of them), the Word Cloud below was generated.

The relevant words that appear most often in our reviews are great, friendly, and family. It came as no surprise that our names were used frequently enough to appear; it illustrates the importance customers place on the personal relationship they form with us. Other frequent words like helpful, fun, and beautiful reinforce

A Word Cloud created from our reviews.

everything we've discussed so far about what agritourists value and remind us where to focus efforts to provide the best customer experience and keep them coming back for more.

Types of Agritourists

Over the last six years of running a pick-your-own, we've observed that agritourists are young, old, single, married, gay, straight, male, female, and everything in between. They represent a wide range of income and education levels. They are racially and culturally diverse, which is plainly apparent at the farm during our season; a regular once mentioned she was out in the field picking, listening to the chatter of other guests, and could hear three languages from where she stood. McKinney, Texas, is not that diverse, y'all.

Scholars of agritourism have grouped participants into three types: purposeful,

Agritourists can be anyone!

recreationist, and apathetic [Barberi 2014]. The purposeful are defined as those who seek to enhance their agricultural understanding or experience through the observation of agricultural processes and participation in hands-on activities. These folks are specifically after a farm-first experience; they want to milk cows, learn how to arrange flowers, find out why their tomatoes didn't produce, and pick their own fruits and vegetables. *These are our true believers.* At Pure Land, we call them regulars.

Regulars are the customers that return to our farm week after week for their routine purchases, often one of several local farm and market stops at which they gather their groceries for the week. They highly value the food we produce, the way we do it, and the personal relationship developed between us. They plan ahead for what they're going to harvest, occasionally picking heavy quantities to put up tomato sauce, pickles, and jelly. They make a beeline for Dad and me when they come out to the farm, phone in hand with pictures of their gardens to ask for advice on this bug or that. They bring along their elderly parents when they're in town because they grew up on a farm, and want them to see yours. They'll take the time to answer common questions posted by the public on your social media, because they know you're busy and just want to help. Best of all, regulars will bring you still-warm homemade brioche and jars of vanilla sugar made with ethically sourced beans from their co-op. They are genuine, generous, supportive people. I might have made three or four relationships like these over the five years we sold at the farmers market, but we now have so many, it pains me to not remember all their names.

The second group, recreationists, are those who seek *any* type of recreation or hospitality activity staged within what appears to be an agricultural setting - in other words, agritainment. These activities include family-friendly events like corn mazes and festivals, fancy farm dinners, stays at a country bed and breakfast, and outdoor recreation like fishing and hiking. Recreationist agritourists are what we refer to at Pure Land as, simply, tourists. The third group, apathetic agritourists, share no preference for a type of agritourism activity, so for our purposes we can lump them in with the tourists. This is your pool of potential new true believers; we want to convert as many of these tourists into regulars as we can.

Tourists generally come to the farm in larger groups and are often families with kids. They may only come once or twice a season but nevertheless make up the majority of the overall demand. The food is mostly incidental to tourists, a by-product of a fun family activity or opportunity to get outside for some exercise and fresh air. If that activity also produces golden social media content, all the better! A good example is the family that visits our farm every summer just one time to pick and take a photo with us to use for the month of June in their family's personalized calendar the following year. Love that!

Both groups are important, but the goal is to convert as many tourists into regulars as possible. In 2022, 36 percent of our guests visited just one time, 28 percent two or three times. Only 13 percent visited more than three times over the eight-week season. The remaining 23 percent had visited the farm in previous years but did not visit once in 2022. The more often we can get folks to come to the farm, the more it becomes embedded in their regular grocery purchasing habits and as a result, sales rise. To make the conversion, give tourists the best local farm experience they could possibly ask for. Make it memorable by offering an organized, friendly experience, making the farm as attractive, safe, and hospitable as it can be, and most of all, knocking their socks off with high-quality produce.

Part 2

PRACTICAL GUIDE TO PICK-YOUR-OWN FARMING

Chapter 3

CONSIDERATIONS BEFORE YOU BEGIN

So, you're thinking about starting a pick-your-own farm. Before you poke a single seed in the ground, there's a lot of work to do to set your business up for success. You're probably expecting to need some dirt and some money, but you may not be aware of aspects such as proximity limitations or particular personality traits that lend themselves better to the pick-your-own model. If you're lucky enough to be starting this journey with some farming equipment and a little piece of property near a metropolitan area, you're way ahead of the game.

The first section will cover general advice for any small farm business just getting started, followed by the specific attributes we've identified as key to the success of a pick-your-own agritourism venture. The goal here is to ensure that all the hard work you're about to undertake leads to success by building your business thoughtfully, from the ground up.

Starting a Farm Business

If you already have an established farm business, you can probably skip this first section, but we wanted to provide some guidance and resources to anyone just starting out. Bear in mind, the information provided here only scratches the surface - whole books have been written on every subject we're merely going to touch on, so I recommend going much deeper than the information provided here. This section is to simply bring awareness to the major considerations when setting up a farm business. Because regulations and policies differ so significantly based on location, much of our advice will be pointing you to the right office or website in your area to consult.

Incorporation

You may or may not choose to incorporate your farm business based on your unique situation and if it would benefit you. If you have personal assets, such as a house, incorporation may protect them from being seized if your farm is sued, for example. However, incorporating will have tax implications, and there are valid reasons for choosing not to do it - talk to a lawyer and tax professional to explore your options before making any decisions.

More important than the specific legal and tax structures you choose is that your farm is established as a legitimate business within that structure from the jump. In other words, do all the things you need to do to operate accordingly inside that structure. If applicable, register your business with the Secretary of State and apply for a tax identification number. Open a bank account solely for the farm and keep your personal finances completely separate. The specific requirements to operate your business legitimately vary based on the legal and tax structure you choose, so consult a lawyer and an accountant to guide you through the process.

Acquiring Land

It doesn't seem reasonable to discuss starting a farm these days without first addressing the growing issue of land access. In 2022, the National Young Farmers Coalition conducted a survey of more than 10,000 farmers and ranchers under 40 and their key finding was that locating affordable land to purchase is the number one biggest challenge for young farmers. According to Regenerate America, farmland in "the countryside" is being gobbled up by corporate developments to the tune of 2,000 acres a day. But even tiny parcels of land are prohibitively expensive in the city, where practicing agriculture is often forbidden anyway. Where, exactly, are we supposed to be farming? And how do we reconcile the demand for locally grown food with the fact that it's increasingly unaffordable to produce it?

There are no easy answers to these questions. One way to help is to lend your voice to groups like the National Young Farmers Coalition, National Sustainable Agriculture Coalition, and American Farmland Trust, who are advocating for policy

Empty lots in the city. Wouldn't it be nice to have a garden here?

changes to increase land access and safeguard farmland from development at the federal, state, and local level. Search out local chapters of these organizations in your area and join them; more than anything, personal stories from farmers are what move the needle towards policy that benefits us all.

Buying Land

If this route is available to you, purchasing land will give you the most freedom and flexibility with your operation. As our good friend and real estate agent always says, "God ain't making any more of it," so land is generally considered a good investment. An internet search of "land access for farmers in [your state]" should reveal government and nonprofit agencies that may be of assistance in financing.

A property that has already been in agricultural use can give you a head start over a piece of raw land, as it will probably have some useful infrastructure in place. Regional websites like New England Farm Finder function as a clearinghouse for farm seekers and farm property holders to connect, featuring properties both for sale or lease; search for a service like this in your area.

Leasing Land

Pick-your-own farming is well-suited for leased property because it requires little infrastructure that can't easily be moved. As you may be hosting hundreds of visitors a day during your season, you'll need to ensure your landlord understands and agrees to the type of farm you plan to run. Being a good neighbor is important in any circumstance but particularly when you're renting and could be uprooted at any time if the neighbors cause a stink.

Drive around the area you want to be in and look for land. Try to meet people in the local agriculture and gardening scene. Many of our customers have offered us space on their land or asked if we knew anyone who might be interested in farming some of it. A lot of properties won't have a "for lease" sign on them, but if you stumble upon a great property and talk to the owners, they may be willing to rent to you. Be sure to negotiate and sign a clear and fair land-use agreement; organizations like Land for Good and the National Agricultural Law Center provide great resources for developing a lease.

Unconventional Land Access

If you want to start an urban farm, find out if your city has a program that allows the use of vacant lots for agricultural production. Some cities and private groups also have "land banks" of acquired abandoned or blighted properties available for rent or purchase specifically for urban agriculture. The nonprofit Trust for Public Land works to protect community access to parks and open spaces including land used for agriculture. Open space and easements like those commonly near hospitals, churches, and schools may be available if you inquire.

Regardless of how you come to acquire land, don't pounce on the first property in your price range without answering some questions.

Physical attributes:

- Is the soil suitable for growing? Have a soil test sent for analysis so you get a general idea of its composition and quality, and to ensure nothing noxious has been used on the land that would preclude fruit and vegetable production. If the land is already in agriculture, ask for the history of what has been produced on it. Poor soil with little organic matter and nutrients shouldn't necessarily be a deal-breaker, just know that it may take several years and a significant investment in practices and amendments to become truly productive.
- Does the general topography work for the kind of farm you want to build? Does the property drain well? Is it in or out of floodplain? Well-draining soil is critical for healthy soil and crops, and the dirt work required to mitigate stagnation and change the terrain can be costly and involve permits.
- Is there a natural water source for irrigation? Have a water test performed to ensure it's suitable for food crops. Treated water (in other words, what comes out of the tap) is better than no water, but it will save money and be more sustainable to irrigate using a natural water source.
- Is it in a location the public will be willing to visit? Not just in good proximity to a metropolitan area which we'll discuss later, but also in a place people will feel safe bringing their kids? Very cheap land in a sketchy area is probably not a great spot for a you-pick.

Families with children will make up a significant proportion of visitors.

Legal attributes:

- Is the property in the city or the county? In other words, who are the governing bodies with jurisdiction over the property?
- What uses is the property zoned for? Can it be used for agriculture?
- What permits, licenses, and infrastructure are necessary in this location to operate a pick-your-own farm legally? Are on-farm sales permitted?
- If you're purchasing the land, how has the property been classified historically with respect to property taxes?

Zoning

Are you legally allowed to operate a farm on the property? Unfortunately, I can't answer that question. Every country, state, county, and city makes their own rules, and crossing the line from one jurisdiction to another can mean a whole different ballgame.

Local governments use property zoning as a tool to shape the community how they want to see it grow and develop. City leaders aren't keen on having a factory churning out smoke next to their entertainment district, or a smelly pig farm in the middle of a new housing development. They want a city layout that feels like it makes sense. An attractive, well-run urban farm is desirable to many residents, but cities are risk-averse and loathe to invite any business that could become a potential nuisance.

The "zone" designation indicates the permitted and prohibited uses of that particular property; some examples of zones are residential, commercial, recreational, and, consequential for our purposes, agricultural. Zones may be set by either the city or the county government in some areas, depending on where you are, so the first order of business is figuring out who has jurisdiction over your property. In Texas, counties have no zoning authority. Our farm lies in an unincorporated part of Collin County, so we are blessed not to have to worry about it. But if your property is zoned, you need to know what it's zoned for and what its resulting permitted uses are.

Demand for locally grown food has led many municipalities to pass provisions allowing production on properties not zoned for agriculture. You can find the permitted uses in your city's zoning code, usually available online at your city or county's website. Residential zones in particular often have many restrictions prohibiting agricultural operations to protect residents from the noise and odors that come along with farming. If you are located in a city or subdivision it is also possible that a Home Owners Association (HOA) may also impose restrictions on some activities. Should farming be prohibited on your property, there is a process to lobby the Planning and Zoning Commission to change the designation or create an exemption or special use rule via a zoning code amendment. The place to start would be consulting a land-use attorney.

Dad and I are real squares - honestly, rule followers to a fault - but some rules are just stupid. Even if you are permitted to farm on your property, make sure you're also permitted to make on-site sales of produce. A good example of some legal nuances that can have a significant effect on your operation can be found in the city of Austin's Urban Agriculture Code. Adopted in 2013, the Code defines Urban Farms as for-profit agricultural use on properties between one and five acres in size, and Market Gardens as those smaller than one acre. These operations are permitted on any property zoned as Single Family Residential. However, a Market Garden may only employ a single full time employee, and allow sales to no more than three customers per day. *Three customers per day!* Of course, it could be worse - in Dallas, produce sales may not be made on-site at all at any urban farm in a residential zone, making pick-your-own downright impossible in a lot of places.

Even if you may technically operate an urban farm, you may not be able to do everything you want as far as the farming itself is concerned. Cities don't like food scraps in the compost pile that may attract rats, so you're likely to face some challenges there. Permission to build accessory structures can also be tightly regulated; an urban farmer we know uprooted his whole operation and moved out into the county because the city wouldn't let him build a greenhouse. We've never had a greenhouse or hoophouse, but if you live somewhere that you can't farm without one, that's something to consider. Needless to say, if you wish to have chickens or livestock, that's a whole other can of worms. Agritourism also requires access to parking, which is bound to be tougher to find in the city. And, lastly, the Fire Marshall or local equivalent will probably want to get involved.

Property Taxes

Before committing to this endeavor, it's important to consider the whole picture of what operating the farm is going to cost. Here, I'm talking beyond the farming equipment, labor, and consumables. If you own your land, one of the biggest bills you may face every year is property tax. If you rent, the burden of payment isn't going to fall directly on you but on your landlord, and may be a significant factor in determining your rent. It's best to figure out what the tax laws are on a property before you purchase or rent land so that you're prepared for the bill.

Property values for taxation purposes are typically established by an appraisal authority. In Texas this is done at the county level and is based on the market value of the property. In the case of land being used for qualified agricultural purposes, the appraised value may be reduced to the "agricultural value" by using measured standards for the output of agricultural products that could potentially be raised on that particular piece of land, the "productivity value." The agricultural value is usually far lower than the market value of the property - consider the difference in dollar amount between the revenue from selling ten acres of wheat versus ten acres of houses. In the 1960s, new developments of sprawling suburbs caused the market value of open farmland to skyrocket. Farmers were forced to sell because they couldn't afford the suddenly massive tax bills, so many states codified "special use valuations," some of which allowed properties kept in agriculture or open space from being taxed at the market value. This is in essence what an agriculture (or "ag") valuation is, often incorrectly also called an "ag exemption" - there is no exemption from the taxes, they're just quantified in a different way that typically results in a lower valuation. The difference in tax burden can easily add up to tens of thousands of dollars a year, a make-or-break situation for many small farms.

The criteria to qualify for special use valuations can be found in your local tax code, and because every appraisal district can set their own, they can differ significantly, particularly when it comes to urban versus rural areas. In some districts, a

Urban sprawl can affect property taxes in the country.

piece of land must be a minimum of 10 acres and have been in active agricultural use the past five years to qualify for agricultural valuation, immediately disqualifying most urban farms just by size. In Texas, an additional, broader Open Space Valuation was passed in 1978. Unfortunately, explicit language describing small, diversified operations like ours is not included in the definition of "agriculture," so many small farms fight to qualify and are typically unsuccessful. Some cities still outright prohibit farms that operate within their limits from qualifying for an ag valuation. Urban farm advocates in cities like Austin have been working for years to have language that describes urban farms added to the local tax code without success.

Income Taxes

The way the operating results of your organization are reported to state and federal tax authorities will depend on how you choose to set up your farm business. Many small farmers choose to operate as "sole proprietors," in which case the business results are simply incorporated into their personal tax returns. If the farm business is established as a limited liability company (LLC) or corporation the entity would obtain a tax identification number of its own and file its own tax returns. The profits, losses, etc. of the farm would then flow to the members or shareholders as is set out in the bylaws of the business. Farm results are typically reported in the IRS Schedule F. If you want to apply for USDA programs like subsidized crop insurance or NRCS grants, the Schedule F is the document they will require for reporting financial results.

Food Safety Laws

At Pure Land Farm, we are what I consider an entry level agritourism operation in terms of the risk of illness or injury to participants. The farm business we run is the simplest, cheapest, safest form of agritourism. Our only product is unprocessed, raw fruits and vegetables; we don't have livestock of any kind, we don't manufacture value-added goods, and we don't sell our products to anyone other than the direct consumer, all of which diminish the risk of contamination and subsequent widespread food-borne illness. But even a product as simple and unprocessed as raw vegetables is subject to federal and local food safety laws.

On the local level, as always, seek out the specific regulations in your area starting with your state's Department of Agriculture. Typically, the act of selling whole, unprocessed self-grown fruits and vegetables on the scale of a small farm does not require any permitting from the state.

Federally, the Food and Drug Administration (FDA) administers the Food Safety Modernization Act (FSMA), passed into US law in 2011. The goal of FSMA is to prevent food-borne illness by proactively preventing contamination rather than merely responding to incidents after the fact. Within FSMA, the Produce Safety Rule regulates standards for growing, harvesting, processing, packing, and holding produce for human consumption. The vast majority of contaminations take place in processing facilities where produce from many factory farms is amassed, sorted, packaged, and shipped to outlets across the country. That process is very different from what happens on small farms that only sell their products locally. For these small farms, many of the ostensibly well-meaning requirements put forth in the name of food safety by FSMA are unnecessary, onerous, and prohibitively expensive.

To accommodate small farms that sell directly to the consumer and in their local communities – what the government calls "qualified end-users" – Congress established some exemptions that the FDA then included in the Produce Safety Rule. Follow the flow charts prepared by national advocacy group Farm and Ranch Freedom Alliance to see the various conditions for exemptions. Our operation, Pure Land, falls into a qualified exemption under the Tester Amendment. With effective bookkeeping, providing the required records to the FDA upon request is almost effortless.

Laws are living documents that are ever-changing; it's important to stay proactive about ensuring your farm business operates in compliance with the legal requirements of the day. Get involved locally with the farm and food policy advocacy groups in your area, like Farm and Ranch Freedom Alliance, which was instrumental in pushing Congress to adopt qualified exemptions to the Produce Safety Rule. These groups are an invaluable resource to keep you informed on the current reg-

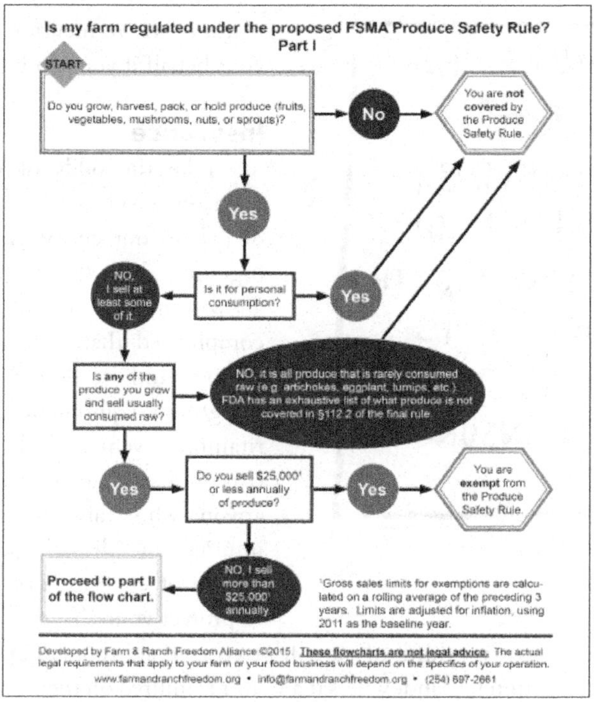

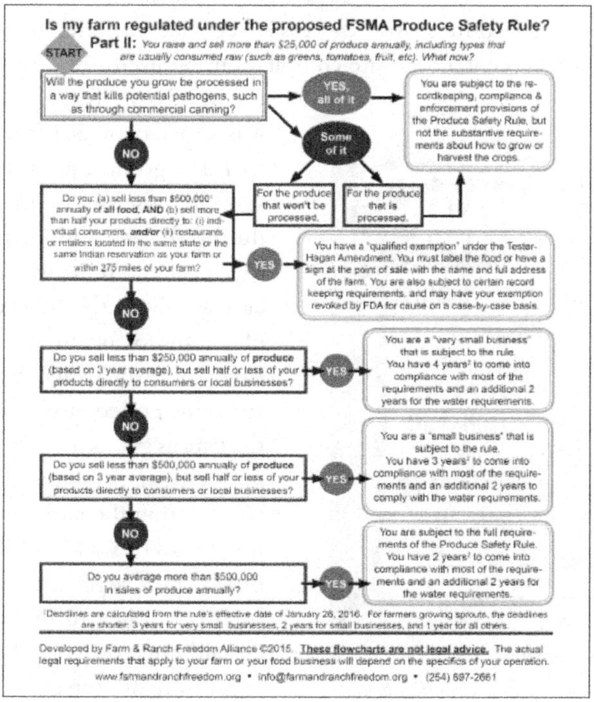

Reprinted with permission from Farm and Ranch Freedom Alliance.

Thy small farm shalt:

BE COVERED BY A MINIMUM OF $250,000 WORTH ($1,000,000 IS BETTER) OF LIABILITY INSURANCE.

ulations, not to mention advocate on your behalf if you run into trouble.

Insurance

Like the odds of getting sick from our produce, there is not a lot of risk to our customers of becoming injured at the farm. We don't allow visitors to use any tools more complicated than a hand shovel or scissors, and we don't have rides or playground equipment like many agritainment venues. Still, guests could trip on a stake, or step into a hole - anyone who's raised a toddler knows injuries can still happen in the most benign of circumstances, so it's best to protect yourself.

Since 2004, some two thirds of states have enacted agritourism immunity laws to offer some protection for landowners if a visitor is injured on their property while engaging in an agritourism activity. The National Agricultural Law Center has compiled these on their website so you can easily see what the laws are in your state (https://nationalaglawcenter.org/state-compilations/agritourism/). In our neck of the woods, the Texas Agritourism Act was enacted in 2015 and stipulates that an agritourism entity is not responsible for any injuries that occur during the activity provided they either post the provided warning sign clearly where the activity is taking place, and/or obtain a signed waiver from the participants. We have two of the signs posted at our farm, one on the gate as guests drive in and another on the barn facing the check-in area. Considering the fairly low risk of injury posed by walking around rows of dirt to pick vegetables, we have not been compelled to implement waivers, although they could offer a greater measure of protection.

Texas Agritourism Act liability warning sign on our barn.

Protection is limited; even with signs posted and waivers signed by every guest, there are situations the Texas Agritourism Act won't cover, so be sure to understand the protection these immunity laws offer in your area. In Texas, these exceptions are:

1) When the employee of the agritourism activity is injured *(you need to provide your own worker's comp)*
2) If the injury was caused by negligence on the part of the agritourism entity *(you need to provide a safe environment and activity)*
3) If the agritourism entity knew an animal included in the activity was or could be dangerous and did not inform the participant *(you may not allow any animals whose behavior could be dangerous to be part of the activity)*
4) If the injury was caused by lack of training of the agritourism entity's employee *(you must train your employees to keep your guests safe)*
5) If the injury was caused directly by the agritourism entity *(you may not punch guests)*

Despite these protections and even if you are truly not at fault, someone can still try to sue you if they are injured on your farm. This can cost a lot of money and time to fight, so agritourism entities are strongly encouraged to purchase their own liability insurance. I wish it had actually been that easy - the problem we ran into for years was getting insurance companies to understand what our business is. As the agritourism industry and the demand for this type of coverage has grown, more insurance companies seem willing to engage. We now maintain a typical farm and ranch policy with $1,000,000 in liability coverage, plus an additional, very affordable premium specifically covering the pickers.

Bookkeeping

An early investment in learning how to painlessly and efficiently keep your books is time well spent. If this is going to be your business and you actually want to take money home, you have to track this stuff, and accounting software is the best way to do so.

At minimum, using accounting software will make filing your Schedule F with the IRS easier, but you'll find there are lots of places you can use the data it captures when you *have* it. Farms that are Qualified Exempt from FSMA are required to keep certain records that you must be able to produce if questioned by the FDA, and yes, we've been asked for them. If you're going to seek out loans or participate in federal programs like Whole Farm Revenue Protection, you'll need to provide copious amounts of historical data, and having a report at your fingertips because your books are complete and up to date will make that effort much easier. The best thing efficient record-keeping has enabled us to do is keep a vice grip on our production numbers to closely monitor how interventions are affecting the crops and to forecast future yields.

We are partial to QuickBooks Online for our bookkeeping which syncs with our bank accounts, integrates seamlessly with a point of sale system, and has a handy app for easy access on our phones. But it's not perfect, and there are many other options out there. There are farm-focused accounting programs like PcMars and CenterPoint Accounting for Agriculture, but an added perk of using QuickBooks or another of the popular, general business software packages is that accounting and tax professionals are more likely to be familiar with them.

Set aside time to reconcile your books and keep them up to date. Implement good habits early on; an example is whenever I purchase anything from a store, I snap and upload the receipt with my phone into the Quickbooks app before I start my car. Doing so means I no longer have to scramble around under the seats three weeks later for that dang receipt. As they say, if you stay ready, you ain't gotta get ready.

Essentials Components of a Pick-Your-Own Farm

We've identified five essential components of a successful pick-your-own farm. These tangible and intangible attributes work together to create the kind of place people want to visit with their families to pick vegetables.

> Thy small farm shalt:
>
> BE LOCATED ON A HARD-SURFACED ROAD WITHIN A RADIUS OF 40 MILES OF A POPULATION CENTER, WITH WELL-DRAINED SOIL AND AN EXCELLENT SOURCE OF WATER.

Access to land near a metropolitan or tourist area relative in size to your farm's scale

No one wants to drive three hours to a farm in the middle of nowhere.

At a market we used to go to, there was a farmer who would wake up at 3:30am to pack his truck and make the three-hour drive into Dallas every Saturday morning. Even back then, I wondered how on earth he made that work financially (I suspect, he didn't).

We surveyed our guests in 2022 to ask how far they drive to visit us, and just over half answered that they travel 30 miles or less. About a third travel fewer than ten miles. I don't necessarily think being out in the sticks completely precludes you from being successful, but you may need to offer more goods or entertainment to convince people it's worth the drive. You could host food trucks or craft and artisan vendors like a farmers market. If you have animals, you could make them part of the show, or hire a mobile petting zoo. Hay mazes and inflatable slides are veering way into agritainment territory, but you gotta do what you gotta do to get people to see the value in paying you a visit if they aren't as moved by the promise of fresh vegetables as I am. But I'd only consider those add-ons if you're already established on a property further than 30 or 40 miles out of town; if you're just starting out, get as close in as you can. Managing a farm store or multiple vendors and activities is distracting, and the focus should be kept on the far more profitable pick-your-own element.

It does not take a lot of land to do what we do. We're actively growing on about four acres, with an additional two for parking and guest areas. We're generous with spacing because we have plenty of it, but the whole operation could be condensed without compromising the experience. We're exactly 43 miles north of downtown Dallas, with dense suburbs filling the entire route between us. Our small city of McKinney itself has a population of just over 200,000 and is growing by the minute; many farms once considered rural are finding themselves suddenly rather suburban.

I specify that the metropolitan area should be of relative size to your farm's scale because if you're planning to have a little bitty half-acre pick-your-own, you likely won't need a pool of a million people to get it all picked. You'll probably be fine near a relatively small city. But if you're trying to rope the moon with a twenty-acre operation, you better have *a real* big city nearby because it's going to take a lot of customers to pick it.

Startup capital for a robust, resilient farming operation

No one wants to visit a crappy operation.

There are so many different ways to farm. If you're just starting out and don't have a clear vision of how you want to grow, visit lots of different kinds of farms to see what kind of farming system speaks to you. A good place to start is Josh Volk's excellent book, *Compact Farms: 15 Proven Plans for Market Farms on 5 Acres or Less.* Then, figure out what you'll need in startup funds to get it off the ground. A strawberry pick-your-own in nearby Greenville, Texas is operated entirely in huge greenhouses with the strawberry plants tucked neatly into three levels of narrow white gutters. Ripe strawberries splay over the sides, making for a very easy (if not particularly natural) picking experience, and people *love it.* The line to get in consistently outmeasures the day's yield of ripe strawberries. The operation clearly made a big investment upfront in infrastructure between the greenhouses, gutters, and heaters.

It will be hard if you're new to farming, but attempt to quantify the expected costs before you begin so you don't end up broke half-way through the season and unable to keep going. We're running a pretty typical market garden setup that has one tractor with a front-end loader, forks, and a few other standard implements. We invested in a raised bedder and transplanter our first season, which has been essential to keeping our farming staff down to just two people. We have lots of grass so we bought a riding mower and a small self-propelled mower for tight spots. We irrigate for free from our pond, but have to buy gasoline to power the pump. The consumables we go through include drip tape, plastic mulch, soil amendments,

The transplanter loaded with pepper plants ready to go in the ground.

> Thy small farm shalt:
>
> BE 'WEATHERPROOF,' AT LEAST AS FAR AS POSSIBLE, WITH BOTH DRIP AND SPRINKLER IRRIGATION.

pest and disease control treatments, and seeds, among others.

It's perfectly fine, even smart in many cases, to bootstrap initially and upgrade later, but we really recommend not trying to open without everything needed to run the farm and provide a hospitable place for visitors. We start full-time farming in March for our summer picking season but don't start making money until May; it's essential to manage funds to keep the farm going before the harvest comes and cash flow starts. It would be a huge bummer to get to opening day and realize there's no money left for a couple of tents. Or there could be an unusually late frost in the forecast, and everything tender might die without a couple grand to spend on frost cloth.

A hospitable climate during your production season

No one wants to be outside when it's too hot, too cold, or hailing.

Some cultures are accustomed to being outside in conditions others find intolerable. One January, we took a trip to Amsterdam. It was about 40 degrees and raining the entire week we were there, but everyone was still outside on their bicycles. That would never fly here in Texas. That summer, we hosted the son of a family friend that happens to live in Amsterdam to work with us on the farm. When the triple digits hit, the boy was *not okay*. Granted, it was a particularly hot summer, but we still had to work through it. We pushed the picking start time an hour earlier to 8am so our last visitors were back in their air-conditioned cars before noon, but by July, we could hardly even get the Texans to come out and pick. Between the extreme heat and increasingly severe weather threatening softball-sized hail and eighty-mile-per-hour winds all summer, it's incredible anyone goes outside at all.

Short of moving, it may take getting creative to carve out a hospitable environment during your picking season. At the end of the 2022 season, I asked our customers what reasons they had for coming to the farm less frequently than they wanted to. Nearly a third listed one of their reasons as it was just too hot. At first we thought, *well, there's not much we can do about that...* But there is. We can work to push our whole season earlier, or at least try to. If we start more crops as transplants rather than direct sow, maybe we could start picking on them a week or two earlier. We could look for varieties that mature faster, too. Last frost permitting, these would contribute to an earlier season, or perhaps a shorter one where we'd

Growing high-quality produce is critical for a successful pick-your-own.

have to let in more pickers on a daily basis (essentially, growing the same amount of vegetables in a shorter time frame).

The ability or deep desire to learn to grow high-quality produce

No one wants to buy crappy vegetables.

A farm could be located on a picturesque property right in the middle of downtown where it's 75 degrees every day, but if the produce sucks, people are not going to buy from them. No one is born knowing how to grow food, just like no one is born with a law degree. You must commit to educating yourself about growing high-quality produce and stay passionate about it.

Fortunately, there are a wide range of resources available to learn good growing skills. For a quick and easy look around hundreds of farming operations, spend a couple hours clicking through YouTube. But remember, *anyone* can start a YouTube channel; there's no guarantee the person on camera knows what they're talking about. Successful farmers like Ben Hartman and Jean-Martin Fortier have created their own online and in-person courses that take you through every step of their production processes and offer direct access to the farmers themselves, a better option for the serious production-minded farmer. AcresUSA, the oldest bookstore in the niche and publisher of this very book, offers in-depth courses by the likes of soil health giants Gary Zimmer and Mark Shepherd through their Eco-Ag U platform. Often you can find esteemed farmers like these at your regional growers associations' annual conferences, which are a great resource for farming advice in your particular area.

Books have been written on every farming technique under the sun. The tattered reference books on our shelves include:

- *The New Organic Grower* and *Four-Season Harvest*, by Eliot Coleman
- *The Lean Farm*, by Ben Hartman
- *No-Till Intensive Vegetable Culture*, by Bryan O'Hara
- *Texas Organic Vegetable Gardening*, by Howard Garrett (find your local organic gardening authority and buy their book)
- *Rodale's Ultimate Encyclopedia of Organic Gardening*, edited by Fern Marshall Bradley, Barbara W. Ellis, and Ellen Phillips
- *Ready Farmer One*, by Diego Footer and Nina Galle
- *The Albrecht Papers*, edited by Charles Walters
- *The Biological Farmer*, by Gary F. Zimmer

And for inspiration:

- *Malabar Farm*, *Out of the Earth*, and others by Louis Bromfield
- *The Unsettling of America* and *The World Ending Fire*, by Wendell Berry

The podcasting boom of the last ten years has produced some great content and is generating more every week. Books are invaluable for their timeless and tested advice, but podcasts offer timeless *and* current information, like discussions on industry trends and responses to global pandemics. We recommend subscribing to:

- Tractor Time, various hosts at AcresUSA
- Farm Small Farm Smart, hosted by Diego Footer
- Growing for Market Podcast, hosted by Andrew Mefferd
- Farmer to Farmer; host Chris Blanchard has sadly passed away but the archives feature virtually every market farmer at the top of the game

Make friends with other farmers, both in your area and on the internet. No one will understand the particular challenges of your growing area like your fellow local growers, so take the time to get to know them and visit their operations if you can. Seek out and introduce yourself to your county extension agents, too. Internet message boards and Facebook groups like the Market Gardening Success Group, with some 96,000 members at the time of this writing, can provide an incredible supply of crowdsourced knowledge (if you can handle the trolls that unfortunately come with all public interactions on the internet).

Things will still go wrong despite your best efforts and it's impossible to control the quality of every piece of produce in the field. There will still be holes in some leaves and peppers with sunburn. The excessive heat and drought conditions in 2022 led to our first ever total tomato crop failure, a humbling experience ten years into our farming careers. It was truly painful to have to reveal the tomato plants to the pickers every day. If, like me, you must search for a silver lining in every situation, crop failures remind the public that farms are living systems and that so much of what happens in nature is beyond our control. More important is how you choose to respond when something bad happens to a crop. For our part, we researched the heck out of what went wrong and armed ourselves with a preventative plan for the following year, which yielded us big juicy slicers once more.

Patience and affection for other people

No one likes feeling unwelcome.

As a friend of mine once put it, many farmers farm because they *don't like people*. I certainly know a few who fit that description, but I also know plenty that do not. Every farmers market seems to have booth or two where the farmers remain seated behind their table the whole time, looking indifferent at best, even hostile towards approaching shoppers. That kind of farmer is probably not cut out for a pick-your-own. Let me put it this way: if talking to people at the farmers market is your least favorite thing about being a farmer, you're probably not going to enjoy agritourism.

Pick-your-own farming requires you to have at minimum a healthy tolerance for the company of other people, because you're going to be dealing with an awful lot of them. While we do spend most of our time on farm work, running a pick-your-own is a heck of a lot different from spending all day, every day on a tractor by yourself and engaging with the public at market just once a week. We actively host pickers some 20 hours a week during the peak of our season; that's a *lot of entertaining*. It may be possible to fake congeniality for a while, but someone who genuinely enjoys the company of other people can really thrive in this line of work. On the flip side, someone who genuinely *dislikes* spending time with other people should seriously think it over before proceeding.

To most people, coming to your farm is a special occasion. It's not a trip to the grocery store, which is considered an actual chore. We face disengaged or crabby staff in grocery stores every day and never think to jump online and write a nasty review. But at the pick-your-own farm, the pressure to please is *on*. You don't have to do jazz hands all day, but you do need to be friendly, welcoming, and engaging to every single person that walks in, one hundred percent of the time. If you're not the chattiest Cathy, hiring extroverts for your seasonal staff can offer some relief, but people always want to talk to The Farmer. Always.

The people that come to see us are the best part.

Understand the powerful position you're in. Remember, the number one reason people have chosen to come to your farm is to support local agriculture and have personal interactions with a farmer. *You can convert these earnest little vegetable pickers into local agriculture fanatics.* This is, in my opinion, the greatest contribution we as farmers who invite the public onto our farms can make in support of the regenerative agriculture movement. Our farm's impact may be tiny by the tons of carbon we sequester or the square yards of native pasture we preserve, but we touch the hearts and minds of thousands of people every summer. *Thousands of people* think a little harder about where their food comes from because of us. Embrace this awesome power and use it for good!

Chapter 4

GROWING STRATEGIES TO MAXIMIZE VARIETY, YIELD, AND PROFIT

After committing to the pick-your-own model, the next step is figuring out what to grow and how to do it. You know, the *farming* part! Anyone who has tried to crop plan knows it can feel overwhelming, but growing for a seasonal agritourism venture is not as hard as growing year-round.

Others in the "six-figure farming club" have developed profitable production techniques that don't incorporate agritourism. At his farm in Quebec, Jean-Martin Fortier operates a biologically intensive system focused on "fast turning" crops that mature quickly, enabling him to increase his revenue by growing more crops over a span of time in the same space. Urban Farmer Curtis Stone uses similar intensive growing techniques at Green City Acres, but he prioritizes the highest profit crops, like salad mix and microgreens, earning more revenue per bed by growing the most expensive crops he can. At Pure Land, we reallocate the labor of picking, prepping, and marketing to agritourists, allowing us to increase our revenue by selling much more than we could ever take to market ourselves.

This chapter is not going to offer gardening advice exactly, as every grower has their own unique struggles with soil fertility, pests, disease, weather, feral hogs, neighbors spraying herbicides, and so on. Rather, we present some strategies for maximizing the yield and variety of your crops to provide the best experience for the pick-your-own agritourist. Clearly identifying the agritourist as the consumer is important; this is *not* a book about farming techniques to make a billion dollars off a tenth of an acre. The product offered by the pick-your-own farmer is substantively different. We sell a fun, safe, educational, family-friendly experience incorporating exercise, nature, and food.

We generally think of this endeavor as slightly heavier on tourism than agriculture but above all else, you *must* have healthy, attractive plants and high-quality food to pick. Ultimately, the majority of the revenue comes from the number of pounds of produce that cross the scale. The guests may have a good time, but if there are caterpillars in the tomatoes and the beets are cracked, they're less likely to pick much or come back. The same is true if they have to wade through knee-high weeds to pick and wake up covered in chigger bites the next day.

There's a lot to consider in a production plan to ensure that guests have a good time while maximizing the farm's earning potential.

Our Farm And Growing Philosophy

Pure Land Farm is located on our 28-acre property on the edge of McKinney, Texas, half an hour north of Dallas. The rich black clay of the Blackland Prairie has a long agricultural history of onions, melons, cotton, grains, corn, and grazing. The property didn't even have an address when we bought it, carved off a larger plot of land by a retired and downsizing neighbor. It was a blank slate with no structures besides a hundred-year-old tin barn. A gentle slope drops about forty feet in elevation from the surrounding farmland to Honey Creek. The first thing we did was terrace the slope into three flat fields of about three quarters of an acre each, totaling a little over two acres in which we rotate annuals. A three-acre blackberry field lies in the rich floodplain at the bottom of the slope, near the creek. We dug a spring-fed pond from which we irrigate through drip tape in the annuals and drip lines in the blackberries. We added electricity and city water about six or seven years in, which has been very nice, but we got by without it. The black Houston clay in which we grow, while mineral rich, has a high calcium content and is extremely sensitive to moisture. The window between being a sticky, unworkable mess and a dry slab of concrete was very narrow for the first several years. With consistent amendments, cover crops, and additions of organic matter it is now much more forgiving and has a pH in the 7.2-7.5 range, down significantly from the original 8.0-8.2.

We have no livestock on the property for a simple reason: no one lives there. While many aspects of this undertaking are a labor of love, it's not a home-

Our view to the northwest.

Blackberry rows.

stead for us. It's where our business is located, and there is nothing about the property that would prevent us from picking up, moving elsewhere, and doing the exact same thing again. It may be a homestead one day, but for now, we choose to focus on the produce with minimal distractions. We mimic the contributions of livestock as best we can through compost, diverse cover crops, and manual inoculations of microbes.

Our general growing philosophy is to use only regenerative inputs, avoid chemical pesticides and herbicides, cover crop extensively, and generally be as gentle and holistic as possible. One exception is the use of minimal tillage and plastic mulch on many of the raised beds. The farm was previously in pasture and the coastal bermuda grass, johnson grass, and yellow nutsedge is extensive and virtually impossible to eliminate. In addition to a measure of grass control, the use of black mulch to warm the soil for heat-loving plants in early spring or white mulch to cool the soil for crops like onions and beets later in the season helps us to push their harvest periods into our picking season window. For these reasons we invested very early on in a raised bed and plastic mulch layer, which has been indispensable to achieving our goals. This piece of equipment attaches to the three-point hitch on the back of the tractor. When pulled, it gathers and packs loosened soil into raised beds, allowing you to also bury irrigation lines and/or wrap the beds with a layer of plastic mulch if you like. We alleviate the damage caused by an annual pass with the tiller by growing produce for only one season, then cover cropping the other eight months of the year. There are so many ways to farm; don't let perfect be the enemy of good. Every year, we strive to get closer to a fully regenerative system.

Deciding What to Grow

Growing for three or four seasons of farmers markets is quite different in practice from growing for a pick-your-own farm and requires a different strategy. The market gardener aims to build up a following at the farmers market by showing up in the same spot with consistent products, week after week. They bring a quantity of produce calculated by knowing how much they can usually sell and try to attract shoppers with something new on the table as the seasons change. A farm that grows year-round does not maximize production on all their land at one time; usually some plots are in production, some are in cover crop, and some are sitting fallow until it's time to plant.

The pick-your-own farmer, on the other hand, should aim to produce a wide variety of crops on as much of their space as possible that can be harvested at the same time. In our experience, guests will pick from as many crops as they are offered, so it is wise to offer as much at once as you can. This is a big change in mentality for the former market gardener, and we got very nervous when we realized just how much we were planting that would be ready all at once. *What are we going to do with all this food if people don't come pick it?*

In planning your growing season, consider that the goal is to cultivate a mix of plants that leave some produce from every crop ready to pick for the last guest of the day without much more on the plants. In other words, the last picker of the day should have the same selection to pick from as the first, but the plants should be picked clean by the end. This is what creates a quality picking experience – no one wants to arrive for the last time slot of the day to find only cherry tomatoes left in the field and no melons, cucumbers, or squash. It's harder than it sounds, but well worth trying to earn happy pickers and good reviews, which we'll talk about much more in Chapter 6.

Conduct a Survey

The first thing we did when considering the transition to a you-pick farm was conduct a survey, easily created online with a free survey app and shared through our social media. (You do have social media accounts, right? You'd better.) We had established a decent following at the farmers market, so I listed out the crops we grew and presented them to our customers with the simple question: What would you like to pick this year at Pure Land Farm? I also included a write-in field to see if there was interest in picking something we weren't growing.

Most of the results were as we expected, with strawberries, peaches, and blueberries easily topping the write-in list. The low popularity of turnips didn't surprise me, but I had been selling thirty or forty pounds of kale during its season every Saturday morning for several years. Only one third of respondents said they

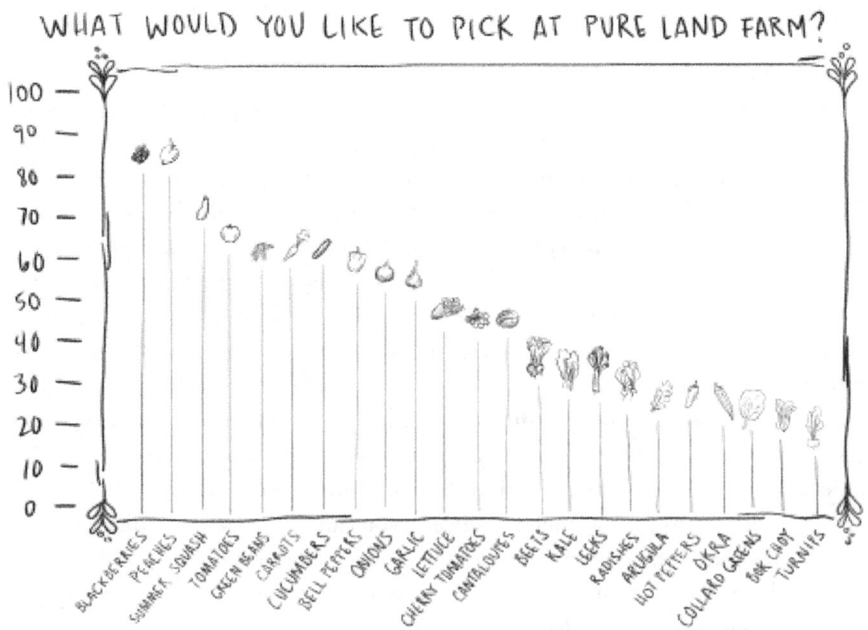

Percentage of people interested in picking each crop at Pure Land.

were interested in picking it. *How could that be?* I planted the kale anyway. Guess what? *They didn't pick it.* And anyone who did would clip four or five leaves, hardly enough for a sad little side dish and amounting to one or two dollars at most. I was so stubborn about it I planted it the second year too, but it was not an aberration. Adding the picking aspect to selling kale changed everything, and no amount of cute how-to-pick signage changed it.

Remember that our results may not be applicable to your market. It is imperative that you find out what *your* customers want to pick, and that you consider a survey to be a jumping off point, not a direct prescription. While our survey accurately captured the lack of interest in picking kale, it didn't tell me that the one third of pickers looking for beets were *fanatical*, and we would struggle to keep up with the volume they demanded for the first couple of years.

It may be tempting, particularly if you are a new grower, to list out every possible crop that grows in your zone on the survey and then plant the top ten from the survey, but be cautious not to put too much money and effort into a crop without knowing how it will perform. There are some crops highly in demand that you'll be unable to produce because of your hardiness zone, soil, unique pest and disease pressure, infrastructure limitations, and so on. Strawberries were the number one write-in suggestion on the survey which led to two fairly disastrous attempts to grow them before we surrendered, but we didn't just scrap all our other crops to plant the whole farm in strawberries because they appeared to be in high demand. We planted a patch of reasonable size to see how they would do, and the financial impact of their failure was largely washed out by the successes of our old standbys.

After getting some idea of what the locals are interested in picking, the next step is to cross reference the results with what you know grows well in your soil and climate. Crops that fall into the middle of the Venn diagram, ones that grow well *and* people want to pick, should be the focus. The survey may also indicate crops to consider scaling up, which is how we ended up with three acres of thornless blackberries.

A success story from this survey analysis for us is summer squash, which we grew well the first year we started farming, then terribly due to pests and disease for

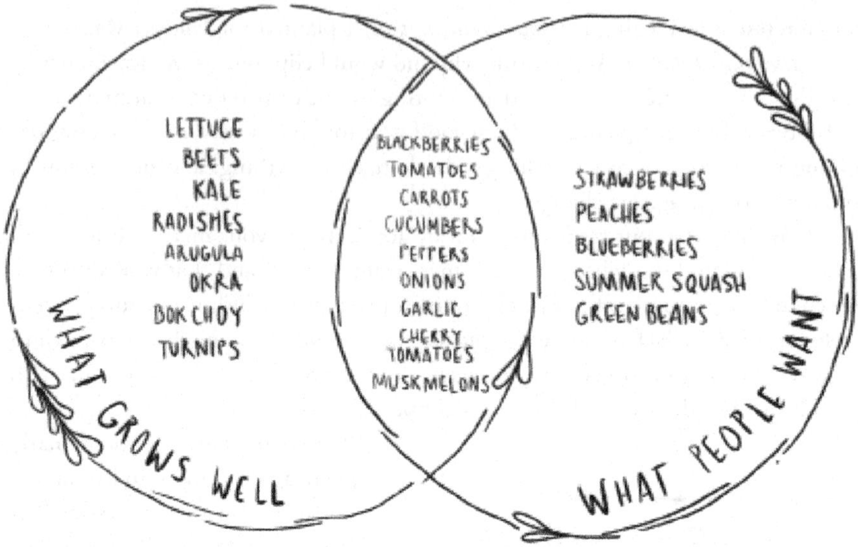

What grows well versus what people want to pick.

the next two until we gave up on it. Seeing the demand from the survey encouraged us to give it another try, and we finally struck a pest control regimen (insect netting, nematodes, and insecticidal soap) that worked for us. I can't say the same for any other crop in that category, unfortunately.

If a crop grows really well for you but the locals say they *don't* want to pick it, plan to scale way back or consider alternative ways to sell it, such as to restaurants or to other farmers who will take it to market. I'm not suggesting to automatically give it up entirely, because morale is important, and it's good for morale to grow things that grow well. But remember, it's creating *more work* to have to pick and market that crop; almost always, the better choice is to let it go and plant something people want to pick in its spot. (Grow a little patch for yourself, visitors probably won't touch it.) Needless to say, if a crop doesn't grow well and people don't want to pick it, drop it from the growing plan without hesitation.

Shorten the Season, Widen the Variety

The more condensed your season, the better. Condensing your season means utilizing as much of your space as possible to grow a wide selection of crops that will be available to pick at the same time, eliminating crops that produce earlier or later. Identify the window where the most in-demand crops grow in your area, and that is your growing season. For us, this is the period between last frost at the end of March and when production drops off due to the heat in late July. This makes our picking season May, June, and July. Your window may be longer or shorter, and earlier or later.

Approximate harvest windows at Pure Land Farm.

It's important to offer a wide enough variety to attract most consumers, which should be attainable for any decent grower in most climates. I would wager that every market gardener has experienced some version of, "Don't you have any tomatoes yet?" in early spring. The consumer doesn't care that it's March; they're used to year-round availability of every type of vegetable grown around the globe at their grocery store. A table at the farmers market in April with a bounty of beautiful early season greens will appeal to some people, but not everyone. In the same way, a pick-your-own in April with only greens to offer for picking will leave many disappointed. But if it's May, and there are late-season greens to pick along with carrots, beets, garlic, onions, and an extra early blackberry, many more people will be excited to visit. Plan your crops with this in mind and wait to open until you can offer an appreciably wide variety as it will make for a better guest experience.

Certain crops can attract plenty of visitors all by themselves, such as blackberries, strawberries, peaches, and apples. If you can make a living growing just one kind of thing and there is sufficient demand for it, you may want to go for it - it certainly simplifies an operation in many ways. In our case, while the blackberries are among our highest revenue crops, the demand is also quite high for the vegetables and we genuinely like growing a wide variety of plants, so we choose to continue to do so. The challenge with increasing crop diversity comes in ensuring that the different nutritional needs and pest and disease mitigation for each crop can be met. Spraying one crop with one fungicide is easy. Spraying three different crops with three different fungicides takes more time and more knowledge. It's harder to grow twenty crops than it is to grow one, but there is valuable risk mitigation and customer appeal in diversity.

Though it's tempting to add earlier or later crops to lengthen the picking season, we generally recommend against it. As an example, here in Texas, strawberry season is March and April. Many of our high-revenue summer crops are planted during strawberry season, and the summer crops benefit from as much attention as possible to get started on the right foot. Running an open picking season on strawberries (while desperately trying to keep them alive) *and* properly attending the new plantings was a real challenge, and our more important summer crops suffered for it. Without overwintering anything, the only other crops ready in March and April would be greens, spring onions, and green garlic. Had we experienced more success with strawberries, we might have explored overwintering a few other things to be ready at the same time. But those crops would take up space that could have been planted with more profitable summer crops, so it would likely put a damper on our overall revenue potential for the year. At the other end of the season, we've had to give up on okra, which produces abundantly in our area but not until late July and August when it's too hot for visitors.

The pick-your-own production strategy differs from the market garden because we are not necessarily trying to get the maximum revenue per bed possible for the crop; we are trying to get the maximum revenue per bed for the crop *that can be harvested in the picking season window.* For instance, in this system it's better to grow determinate tomatoes rather than indeterminate. Determinate tomatoes, also referred to as bush-type, bear their entire yield in a short period, usually just a few weeks. There are no cultivation techniques that can increase the yield; the number of tomatoes they will produce is "determined" by their genetics. Indeterminate tomatoes, or vining-type, can yield more tomatoes than determinates overall, but do so over a much longer period of time. For a pick-your-own, it is better to grow determinate tomatoes that may produce fifteen pounds of tomatoes per plant over a short period when there is a great variety of other crops at the farm ready to harvest, than to grow indeterminate tomatoes that may produce thirty pounds per plant over four months when you may not have much else producing and it's hard to get people to come out. You may choose to include some crops because they provide variety even if your picking season doesn't fall on their peak; if you can still achieve sufficient revenue per foot, there is no reason to exclude those crops just because they don't reach their maximum possible output.

Above all, a condensed season with a reasonable amount of crop variety improves the guest experience, and that's what keeps them coming back. But shortening the picking season also creates an obvious benefit to the farmer: a shorter season of work. If it was possible to make the same amount of money (or more) in a few months as you could in a year, which would you choose? Then, the way you spend your off season is up to you. When the triple-digit dog days of Texas summer hit, our farm production slumps, and it becomes impossible to keep the produce quality high and the people coming out. That's our annual signal to close.

It is entirely possible to have a fall picking season in our climate, and we are asked about it often, but it wouldn't automatically double our sales. Seasonality dictates that you can't just grow the same thing in the fall as you can in the spring, and most of the crops appropriate for the second picking season here are both lower-revenue and lower-yielding (think, one head of cabbage in the fall that takes up two feet of row and might fetch an ambitious $4, versus a zucchini plant in the summer that produces $25 of squash in the same two feet). In our area, the harvest period for summer crops and the growing period for fall crops can significantly overlap, so we would likely have to carve space away from the more profitable summer planting to grow a good variety in the fall. Even if we did try to squeeze two picking seasons into the same space by growing crops that didn't overlap, it would virtually eliminate meaningful cover cropping since we would be left with only a couple weeks in the hottest part of summer and a few more in the dead of winter to do so. Finally, a big part of why the late spring and summer picking season works so well for us is the alignment with the school year. When school is in session, could we attract enough visitors? Maybe, maybe not. We choose to run the shortest season possible and fully maximize production on our space at the time of year when the weather gets nice, kids are out of school, and everyone seems to remember the farmers market exists. As the adage goes, make hay while the sun shines.

Maximize the Earning Potential

Maximizing the farm's earning potential will come not just from growing what the customers want to pick, but doing so in a way that prioritizes crops that yield the most revenue. Obviously, growing what customers want is the smart thing to do, but do those crops make enough money? Over the first couple of years of you-pick we increased our revenue by increasing the size of our fields, but we did not scale up before ensuring it was not more than we could properly take care of *and* that there was enough demand for it to get picked out. Currently we have maxed out the size of the annual planting we can manage as a twosome, so any changes in revenue are only due to tweaking the crop mix.

It's a concept across many industries but in farming, a "loss leader" is a popular but unprofitable crop grown to attract shoppers in hopes of encouraging the sales of other crops. Sweet corn is a good example; it takes up a tremendous amount of space and is expensive to produce on a small scale but is so popular with

> Thy small farm shalt:
>
> CONSIST OF COMPATIBLE, COMPLEMENTARY CROP COMPONENTS THAT EARN A MINIMUM OF $3,000 PER ACRE ANNUALLY.

consumers that many market gardeners still grow it anyway. At a pick-your-own farm, we have a captive audience. The shopper who has made their way to our farm has already committed to buying what we're selling, so loss leaders are unnecessary and only hurt the bottom line. If we try everything in our farming arsenal but can't make a crop reach our self-imposed minimum revenue target, we stop growing it.

While we are firm believers that necessary costs are to be minimized and unnecessary costs eliminated, we do not generally allocate costs by crop. If it costs too much to grow in terms of inputs or attention (we call these "fussy" plants), we don't grow it anymore. We view costs in a more holistic way; the goal is to maximize the profitability of the entire enterprise, not the individual crops, and one of the essential ways to accomplish this in a diverse pick-your-own operation is to optimize the *mix* of crops. At Pure Land, our costs make up about 35 percent of our annual gross. The bulk of the difference in the cost of producing one crop compared to another is generally not in the growing, but in the harvesting, processing, and marketing end of the business. Carrots must be pulled, washed, and bunched before they can be taken to market with everything else and sold. By allowing visitors to execute these tasks (excuse me - *enjoy these tasks*), the costs of growing carrots are largely equal to other crops and thus, after satisfying enough variety to be appealing to most agritourists, we decide how much of each crop to grow based solely on which crop mix makes the most money.

Experienced market gardeners likely already have some kind of system for estimating yields, which can vary wildly in accuracy depending on the system used and how disciplined they are about keeping up with it. When we sold through the farmers market, we would roughly allocate the day's sales to individual crops by comparing what we brought home to what we took to market, rather than taking on the daunting step of weighing every crop on the way in and on the way out. This, combined with the data from wholesale transactions captured through the invoice process, gave us *some* idea of yield and revenue by crop. Still, we didn't tie it back to how much of that crop had been planted. Decisions to plant more or less of any given crop were largely based on generalizations like "we always sell out of beets," or "we always have lettuce left over."

One of the unforeseen perks of the you-pick model has been the precision with which we are now measuring our yields. We know crop by crop, day by day, every pound picked and sold because everything that leaves the field hits the scale at checkout and is automatically logged in the payment transaction software. Through this data, it's possible to glean all kinds of information. We can adjust variables in cultivation technique and measure changes in yields year on year. We can predict crop availability to our customers during the season because we know from the sales data that the Prime Ark Freedom™ blackberries hit their peak production about seven days after the first berry ripens, the Natchez variety starts ripening about ten days after that, and so on. It also allows us to optimize the mix of crops in the available space to forecast our future revenue, toggling more feet of this and fewer of that to hit the numbers we want. Should you pursue acquiring diversified crop revenue protection insurance such as the Whole Farm Revenue Protection (WFRP) program by USDA, this information will be essential.

Calculating the revenue per unit of area farmed is essential to maximizing the overall revenue. Once you start accurately calculating this information, you will find it incredibly valuable. Some might prefer to use beds, or square feet, or some other unit convenient to your operation; it doesn't matter. Use whatever units work for you. What matters is tying the revenue from each specific crop to some measure of the space utilized to produce it. We do everything in rows of raised beds of approximately the same width, so our measure of choice is revenue per linear foot of row.

Unfortunately, the projected revenue per foot does not exist in a vacuum, and no two years are exactly alike. In our area and across the globe, farmers are experiencing increasing extremes in temperature and severe weather every year. Record lows from Winter Storm Uri and a historically late freeze in the spring of 2021 decimated two thirds of our blackberry crop, destroying their projected revenue per foot for the year. All we can do is work to control what we can to maximize production under whatever the uncontrolled circumstances may be.

Calculating Revenue Per Foot

Simply put, the revenue per foot tells you how much money to expect to make per row foot of each crop based on your historical data. If you don't have any data to work with, start with an estimate and refine the number after your first year. We'll get more in depth about pricing later, but our general suggestion when starting out is to align your prices with the other market gardeners and farmers markets in your area. To calculate the revenue per foot, divide your total revenue from each crop for the season by the row feet that were planted.

Once you have captured a reasonably accurate revenue per foot, the next step is to analyze how to increase it for each crop, particularly those on the low end. Set a threshold of minimum revenue required per foot and take a hard look at those below the line. At Pure Land, any crop producing less than $5 per foot of row is in danger of being dropped from the roster if we can't raise it with one of the strategies

SAMPLE CALCULATIONS OF REVENUE PER FOOT FOR SEVERAL CROPS IN 2023

Crops	Row feet planted	Total harvest LB or EACH	Price	Total season revenue	Revenue per foot
CHERRY TOMATOES	189	1066	$4	$4264	$22.56
CUCUMBERS	732	3738	$2.50	$9345	$12.77
CARROTS	849	2437	$4	$9748	$11.48
FLOWERS	309	1712	$1	$1712	$5.54
MELONS	1417	4442	$1.50	$6663	$4.70

Revenue per foot allows you to compare how much you could earn in one unit of area from different crops.

below. The melons in the chart above, for example, need the revenue per foot to increase for us to keep growing them. When a crop is popular, as is the case with our melons, it's worth going through some troubleshooting before walking away from them entirely.

Increasing the Revenue per Foot

Observe the performances of both the crops and the pickers throughout every season and take notes. Which crops are consistently picked clean? Which crops are most labor intensive or susceptible to problems? What specifically prevented any given crop from yielding a higher revenue per foot?

After making observations, consider applying one or more of the following strategies to each crop to increase the revenue per foot.

OBSERVATIONS OF CROP AND PICKER PERFORMANCE DURING A SAMPLE SEASON

Crops	Revenue per foot	Observations
CHERRY TOMATOES	$22.56	Never picked clean; overripe fruit falls off and is gross
CUCUMBERS	$12.77	Picked clean every pick, but already planted quite intensively
CARROTS	$11.48	Very popular, but pickers need a lot of guidance, and weeding is a ton of labor, so we do not want to scale up any further
FLOWERS	$5.54	Sunflowers were by far most popular; others not picked much
MELONS	$4.70	Picked clean every pick; appears to be enough space between plants to move closer together

Make observations of crop performance and picker behavior.

Increase Yield per Plant

The first approach is one to employ across the board, whether a crop is meeting the required revenue per foot threshold or not. In the best-case production scenario during the best year ever, what is the yield potential per plant or row foot of each crop? When it isn't met, why not?

If your business is selling onions by the pound, you better be trying to grow the biggest onions you possibly can. It helps that agritourists are generally excited by finding and picking large fruit and vegetables. (Interestingly, this is the opposite position generally taken by chefs, which compounded the difficulty we had in making a profit selling to restaurants.) Get back to the basics of good growing; send off a soil test, amend appropriately, and take the non-production time afforded with this model to incorporate cover crops. Diagnose and respond to nutrient deficiencies and be armed to fight the same bugs and diseases that appear every year.

In climates where weather conditions can be harsh and swing dramatically during the season, variety selection can be critical to good production. Over the years we've experimented with growing over four dozen varieties of tomatoes, but we now only grow the same handful every year. Find the best-suited varieties for your unique soil, pest and disease pressure, and microclimate. Every year we still test a few new varieties just to see if they do better than our old standards, and if so, we swap the new ones into the roster.

Optimize Plant Spacing

If a crop is producing well with no signs of stress at the current planting density, consider spacing them closer together. We continue to grow melons because they are highly in demand and grow relatively well, but the revenue per foot the way we originally spaced them was very low. We experimented with moving them closer together which more than doubled the revenue per foot, but they still don't hit the $5 per row foot threshold. Next year, we will move them even closer together and see what happens.

Garlic is an example where although it meets the minimum threshold, the revenue per foot could be improved upon. Historically, we grew garlic spaced in three rows across the top of the bed with about a foot between each bulb. Research and comparison with other growers suggested we could fit a fourth row onto our beds without compromising their final size or quality, and our garlic revenue increased by 25% at the negligible cost of more seed.

Certain crops like beets do well multiplanted, a Dutch technique popularized by Eliot Coleman. Four or five seeds are sown into each cell, then transplanted as a clump rather than as individuals. *This is very appealing when planting 12,000 beets.* Each beet in the clump pushes out against the others as they grow, with no negative impact on the yield. This technique also allows for laying plastic or biodegradable mulch to control weeds, because the clumps are planted in three rows with one foot spacing down the row, rather than broadcast or sown three inches apart across the

Multiplanted golden beets.

whole bed.

Bear in mind when intensifying crop spacing that there is a point of diminishing returns. Plants without adequate spacing can become diseased due to lack of airflow or can scare pickers off, as we observed with a radish crop a few years ago. Sown too thickly in places, pickers were unable to see the individual plants and didn't know how to approach them. In the areas of the row where the radishes were sown less densely, they became picked out. Ultimately, we plowed a lot of radishes under that had become overgrown, wasting space, time, and product.

Improve the Picking Experience

The experience on a pick-your-own farm is what makes the carrots different from carrots at the farmers market or in the grocery store. Providing a safe, easy, fun picking experience will yield higher sales per guest and good reviews, which will bring even more pickers.

Consider this case illustrating the value of the picking experience. Onions are a staple for us, but while great tasting and low maintenance, they were only yielding $3.74 per row foot a few years ago. We were already planting them so densely they practically touched when fully bulbed. We continued to work on our soil health and optimizing our amendments, resulting in decent-sized but not huge onions, so there was some room for improvement there, but not a whole lot. We also always had onions left over at the end of the picking season.

The summer of 2019, we noticed something interesting in the trend of our onion sales. We had always planted short-day onions as suggested for our area, which are nicely bulbed when the picking season starts in mid-May. However, by mid-June, a full month before our picking season ends, the unpicked onions need to be harvested and cured or they'll rot. We pull and cure them ourselves, then display them near the checkout for visitors to add to their baskets. At the beginning of the season when visitors could pick their own onions, we reliably sold about 80 pounds every pick. But later in the season when they were just grabbing cured onions at the checkout, the sales dropped by half. *The picking experience was 50% of the value.* Not only was our picking and curing the onions negatively impacting their sale, but it also created *more work* for us! Talk about a lose-lose situation.

Onions in the process of being harvested.

To increase the number of onions that would be available for guests to pick, we needed onions that matured a little later so they could stay in the ground longer. The following year, we planted half our normal short-day varieties and half similar intermediate-day varieties, and it worked like a charm. The intermediate-day onions were ready a few weeks later, so we concentrated pickers in the short-day varieties first, then let them into the intermediate ones as they matured, leaving hardly any short-day onions left for us to pick and cure when they needed to come out. With that effortless change, the revenue per foot of row is now up to $10.85 and our picking and curing labor is minimal.

Another simple improvement to the picking experience is weeding. Managing weeds and grass is a constant battle, but keeping the beds and pathways clear always leads to bigger sales. Guests do not like wading into knee-deep grass to get to the plants, and they'll give up quickly and move on if they can't figure out which thing to pull to get a carrot.

Increase the Price

Years ago, I asked a fellow grower at the farmers market what he charges for his watermelons. Without a moment's hesitation, he answered, "As much as I possibly can." While I don't subscribe to this philosophy unchecked, he does have a point. Compare prices to other local growers and charge what the market is willing to pay. Remember that during agritourism experiences, customers are not motivated by low prices.

When growing crops that are not usually purchased in bulk, extrapolate a price per pound from the price of the unit typically purchased. Market gardeners in our area usually pop six or seven big jalapenos in a container for display on their table and charge something like $2, giving jalapenos the relatively high price per pound of $8-10. When we first converted to the you-pick operation, I was afraid that kind of price tag would scare people off and reduced the price of my jalapenos to $5 per pound. But the consumer is aware that a pound is *a lot* of jalapenos. Rarely do people check out with more than a handful of them, and paying a dollar or two for that amount is considered reasonable. I raised the price by nearly double with no issues, and now all our prices have been fine tuned to about as high as our market will bear.

Some crops simply grow in units too large for a profitable per pound price to be reasonable to the consumer. For example, we are frequently asked to grow watermelons. As growers know, watermelons take up a lot of space and produce only a couple fruits per plant. They're typically grown many acres at a time, not as a small farm row crop. To reach a decent revenue per foot in our operation, the per pound price would be well beyond what most people would consider reasonable, easily adding up to $30 watermelons. Therefore, we don't grow them. One mitigating tactic for this is to plant varieties that produce a greater number of smaller fruit, which will drop the per-item price into a more palatable range. Or, if you have acres of space and grow amazing watermelons with little maintenance, you might be able to tolerate a much lower revenue per foot and find growing them worth it (good luck explaining how to pick a ripe one, though).

Increase the Number of Pickers

Some crops will be picked clean every pick, and some will seem barely touched. During our first season, we would often find absolutely no peppers or cucumbers left in the field, but the cherry tomatoes were still loaded. To this day, we have never even come close to being picked out of cherry tomatoes. Our first instinct was that, clearly, we were growing too many. But upon reviewing the revenue data, we discovered they are our most lucrative crop. Why would we plant less of our most lucrative crop? To increase our overall revenue, we needed to plant more of them, and increase the number of people picking them. However, it doesn't work to increase the number of pickers just to take care of one crop's abundance; to provide a good experience, a balance must be struck between the crops that produce abundantly and the ones that get picked clean. Bare plants lead to disappointed pickers, so take care to optimize the whole plant mix when increasing the overall number of pickers.

Create Additional Sales Avenues

If you end up with too much ripe produce at once due to an unbalanced crop mix or bumper crop, there are a few ways to deal with it. You can do the picking yourself and set up a farm stand for off-days, sell wholesale, or drop the surplus at the local food pantry. Some farmers use a glut to make shelf-stable value-added products like jam or pickles which can then be sold to guests during picks. However, *nothing* is as profitable as the guest picking and purchasing produce in its unprocessed state.

Let's say there are a lot of ripe blackberries on the plants. Some might assume that because there are "excess" blackberries, it won't cost anything but jars and a couple ingredients to make blackberry jam. But take a closer look. I can get about thirty half-pint jars of old-fashioned, pectin-free jam from ten pounds of blackberries. I can sell them for the relatively high price of $8 per jar, totaling $240 in revenue for the lot.

Labor
- Time picking: 1 hour x $7.25/hr
- Time making jam: 4 hours x $7.25/hr
- = $36.25

Cost of inputs
- Blackberries: $0
- Organic lemons: $3
- Organic sugar: $7
- Labels: $5
- Jars: $25
- = $40.00

Total cost to produce = $76.25

Net profit = $163.75

If I pay myself minimum wage, it takes at least five hours of *not farming* to sell ten pounds of blackberries and net $163.75 in "added value." Subtract the $70 in actual blackberry value, and the profit is down to $93.75. I probably should have been working on the farm those five hours, too; either I need to pay someone else to do the farm work while I'm making jam, or, more likely, I'll just try to play catch-up and get behind.

At our farm, each guest picks an average of 1.4 pounds of blackberries. Having a mere seven extra guests to pick would have sold the same ten pounds of blackberries *with* absolutely no additional time, effort, or expense. To the contrary, the guests would have paid admission and likely picked other things, too, quickly reducing any "added value" to zero or a loss.

With enough manpower, storage, and production, a farm stand could be an effective means of dealing with surplus produce. We lack two of the three, but increasing production is leading us to consider adding manpower and storage. No matter what, pickers do not get every ripe piece of produce out of the field, and we do have customers more interested in the produce than the picking (particularly older folks). If the effort brings in more than it costs to execute, having a staff member harvest, prep, store, and market surplus produce in a farm stand on site may be the next most efficient means of sale.

Going back to our example crops from earlier, let's apply a potential strategy to increase the revenue per foot of each crop.

At the end of every season, revisit these observations and calculate the difference in revenue per foot from the previous year. What worked, and what didn't? What new things went wrong that you weren't prepared for? Then use these strategies again to aim higher every year.

OBSERVATIONS OF CROP AND PICKER PERFORMANCE DURING A SAMPLE SEASON WITH REVENUE-RAISING STRATEGIES APPLIED TO EACH

Crop	Revenue per foot	Observations	Strategy to increase Revenue per foot
CHERRY TOMATOES	$ 22.56	Never picked clean; overripe fruit falls off and is gross	Increase overall number of pickers
CUCUMBERS	$ 12.77	Picked clean every pick, but already planted quite intensively	Increase row feet
CARROTS	$ 11.48	Very popular, but pickers need a lot of guidance, and weeding is a ton of labor	No change
FLOWERS	$ 5.54	Sunflowers were by far most popular; others not picked much	Improve picking experience; Only plant sunflowers
MELONS	$ 4.70	Picked clean every pick; appears to be enough space between plants to move closer together	Intensify planting and/or increase price

Apply strategies to increase the revenue per foot.

Putting it All Together

After settling on what to grow, it's time to lay out the field. Start with a relatively small planting, then scale up year on year. No system becomes more ordered over time without intervention, so keep your farm sized appropriately to your available labor. Many fledgling farmers and gardeners plant far too much, not realizing the amount of attention the crops will take to be productive. Then they run out of time and miss a spray and accidentally let the field go an extra day without watering, then the grass is two feet high and they drop everything to mow, now they've run out of time to plant that next succession, and everything falls apart. For the last couple of years, we have not increased the size of our planting but it yields more because we're optimizing the production and picking efficacy per plant. A smaller number of plants that are well cared for is more productive than a larger number of plants that are neglected. Attempting to run an operation that's too big has negative effects on your:

- Money. Plants that are not properly tended will produce lower yields.
- Time. The to-do list never ends even on the most efficiently run farms. Being unable to complete tasks thoroughly leads to patch jobs that require more time in the long run.
- Mental health. Stress and poor morale from feeling like you're working nonstop but barely keeping your head above water is a recipe for burnout.

The exact amounts of each crop to grow should be set both by the revenue per foot and by qualitative observations gleaned from your previous seasons and your market. What are your most popular crops, or those most popular in your area? What crop do you grow that is impressive but under-picked, one that you feel that you could build a bigger market for? Once you have a season under your belt, and every season thereafter, you will be able to fine tune your fields to meet the unique demands of your market.

As a place to start, this is the current mix of annuals we grow at Pure Land. While differences in climate and local tastes matter, these are basic, bread-and-butter crops that the general public should be familiar with no matter where you are. With these relative amounts of each crop, we are usually able to present the same offerings to the first picker and the last picker of the day while maximizing our potential revenue. Your mileage may vary; it is paramount to tailor your crop mix to the demands of your own market and refine it from season to season. The blackberries and recently added experimental raspberries are excluded from this analysis as they are perennial, so there are few changes that can be made year-on-year except for planting more.

Combine your qualitative observations over the season with the revenue per foot data captured at the end and strategize how to tweak the crop mix for the following year to maximize your earning potential. The most important question we ask for each crop is, was it picked clean every pick? In other words, what did we have too little or too much of, and why?

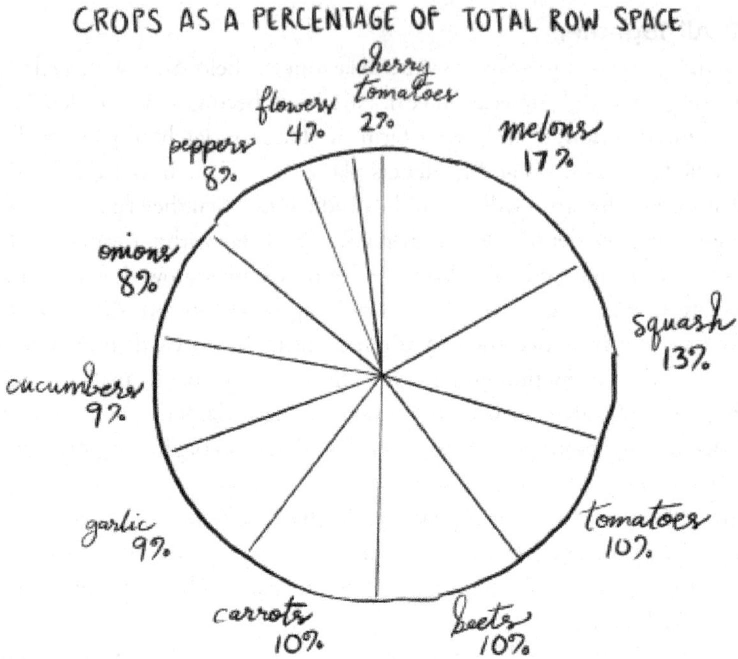

The crop mix at Pure Land Farm in 2023.

Crop	# OF plants
cherry tomatoes	65
tomatoes	306
flowers	457
squash	549
peppers	1019
cucumbers	1082
melons	2095
beets	2951
garlic	3447
onions	10645
carrots	100,000 (broadcast)

Total number of plants of each crop in 2023.

T3

ROW		
1	Raspberries	Raspberries
2	Cantaloupe	Cantaloupe
3	Raspberries	Raspberries
4	Cantaloupe	Cantaloupe
5	Raspberries	Raspberries
6	Raspberries	Raspberries
7	Cantaloupe	Cantaloupe
8	Raspberries	Raspberries
9	Cantaloupe	Cantaloupe
10	Raspberries	Raspberries

T2

ROW		
1	Flowers	onions
2	carrots	onions
3	carrots	onions
4	slicing tomatoes	onions
5	slicing tomatoes	cherry tomatoes
6	slicing tomatoes	slicing tomatoes
7	slicing tomatoes	garlic
8	carrots	garlic
9	carrots	garlic
10	carrots	garlic

T1

ROW		
1	Flowers	beets
2	Hot peppers	beets
3	Bell peppers	beets
4	Bell peppers	beets
5	Bell peppers	beets
6	summer squash	summer squash
7	summer squash	summer squash
8	summer squash	summer squash
9	cucumbers	cucumbers
10	cucumbers	cucumbers

Our farm is terraced into the hillside to create three fields of about ¾ an acre each, referred to as T1, T2, and T3. Each terrace holds about 3,500 linear feet of row spaced six feet apart in 30-inch-wide beds.

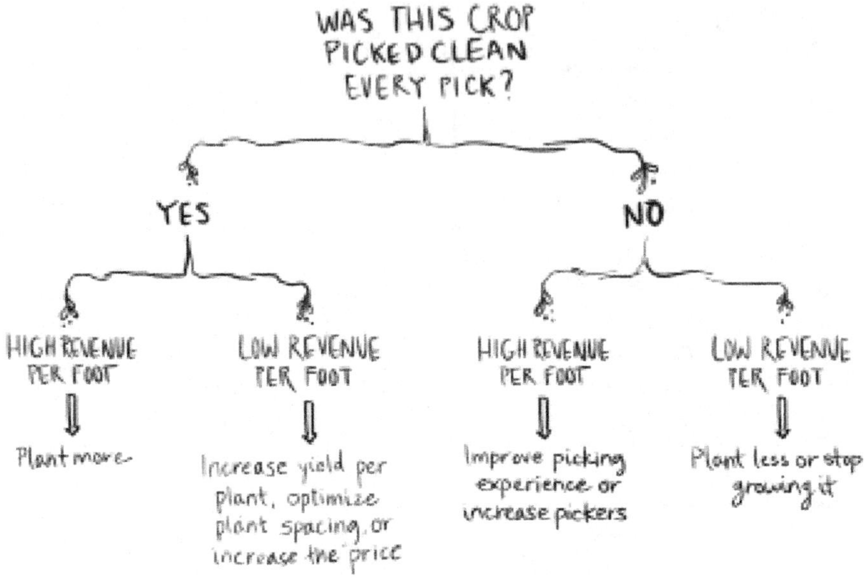

Ask first, was it picked clean? Then, apply strategies to optimize the crop mix and maximize revenue potential.

Crops that are always picked clean and yield a high revenue per foot should be scaled up until the demand is met, observed when the crop is no longer picked clean. On the other side of the flow chart, crops that are not picked clean and yield low revenue per foot get the boot. The only reason to continue growing these would be to bolster the appearance of a wider variety of crops to attract pickers. But don't plant too much, because they are costing money that would have been made from a higher-revenue crop in their place.

If there are crops that are always picked clean but yield low revenue per foot, troubleshoot how to raise it with the strategies described earlier in this chapter. Can you increase the yields with a variety better suited for your environment or with cultivation techniques? The inflation we've all experienced in the years following the COVID-19 pandemic and the associated rise in costs of doing business justify a good look at raising prices across the board, whether for the produce, admission, or both.

Finally, if a high revenue per foot crop is not getting picked clean every week, try to figure out why. There may just not be enough demand, which you can try to stoke by talking it up to pickers and providing recipes. Improving the picking experience can almost always help; has the grass gotten out of control in those rows? Or are they planted too densely for the pickers' comfort? In the case of our cherry tomatoes, the picking experience is as good as we can possibly provide, so increasing the number of pickers is the only solution. But remember, it doesn't work to increase the number of pickers without changing anything else; it's important to increase the planting of the crops that do get picked clean as well, otherwise there will only be cherry tomatoes left for the last pickers of the day, and that's not a good experience.

General Considerations

Many visitors will be experiencing a pick-your-own farm for the very first time when they come to see you. While they're usually excited, they can also feel nervous to do something so new and unfamiliar. Making it as easy as possible for visitors to access the plants and understand what to do will result in happy pickers and bigger sales. Here are some general suggestions and pitfalls to be aware of on the farm production side.

Crops

Everyone is going to see how the sausage gets made.

There is no concealing ugly, dying plants or bug-damaged produce at a pick-your-own. A farm we know always sells their organic sweet corn with the top inch cut off. The consumer doesn't really notice, but anyone who has ever tried to grow organic sweet corn knows why they do this; to remove the fat caterpillar burrowed into the tip of every ear. I am totally in favor of this practice - the consumer never knows the caterpillar even existed, and if any farmer could successfully get rid of corn earworm without genetically modifying the corn, they would! Instead, this perfectly delicious, *perfectly good* sweet corn gets sold and eaten, not wasted because of a minor and virtually unavoidable imperfection. On the pick-your-own farm, this is obviously not something a farmer can pull off. The entire growing process is exposed. If a crop is unappealing if not sorted or prepared in some way for market it should be dropped from the crop mix, or risk bad reviews about the quality of the produce at your farm.

We don't grow corn, but an example we deal with at Pure Land is Golden Nugget cherry tomatoes. This variety produces abundantly so we have grown it since we started farming. The only drawback is one light rain shower and the skin on every ripe Nugget splits. For the farmers market this was no big deal, I would simply pick around them and let the split ones go soft and fall off. All my customers saw were tills of perfect cherry tomatoes on the table. But agritourists are justifiably turned off by a plant full of mushy split tomatoes, and they aren't interested in fishing through them to find the good ones. What we settled on for the sake of maximizing variety and profit is growing far fewer Golden Nuggets, committing time to removing and cleaning up the split ones when it rains, and growing more of the varieties that don't split as readily.

Aim for familiar varieties

When starting out, aim to grow the varieties of each crop most recognizable to the public. Market gardeners often have a roster of fancy and unusual heirloom varieties for restaurants and savvy farmers market shoppers. It's tempting to offer these to the agritourist, but in our experience, they greatly prefer to pick what is familiar. For years we grew several colors of carrots, but it never failed; the orange ones were picked out long before the others. If you dedicate space to a goofy variety that goes unpicked, you've not only lost potential revenue but real money in inputs and amendments. Resist the urge to plant what you think your visitors should be picking in favor of letting them guide you to plant what they want.

Aim for familiar varieties.

Stagger harvests by planting varieties with different days to maturity

To prolong the harvest period of a crop, try planting varieties that differ in days to maturity rather than staggering the planting time of the same variety. It's more efficient to plant a full crop at once instead of doing two or more rounds of seed starting. Look for varieties of the same crop that have a decent spread in time to harvest, and plant them at the same time. Just be sure the later variety doesn't end up too late and is only hitting its stride when everything else has finished producing.

Use crop supports to improve the picking experience

Pickers generally like plants that are upright, where the produce is exposed and easy to reach. Consider the tomato picking experience. We always start double the tomato plants we need as insurance against late frosts, then give away and/or compost the ones we don't end up needing. In 2020, we were faced with unprecedented demand to come pick during the COVID-19 crisis, so we decided to prep a few more rows and plant the backup tomatoes. We grow determinate tomatoes and only had enough cages for our planned number of plants, not for the backups, so we decided to let them grow without support and see what happened. They produced just fine, but it was obvious very quickly that most people are not comfortable reaching into a bushy, gangly tomato plant sprawled across the dirt to fish out the tomatoes buried within. The regular plantings, which were caged and standing up clean off the dirt, were picked much more thoroughly.

For safety, be careful not to create trip hazards with trellises and make sure all crop supports are free from jagged or sharp edges.

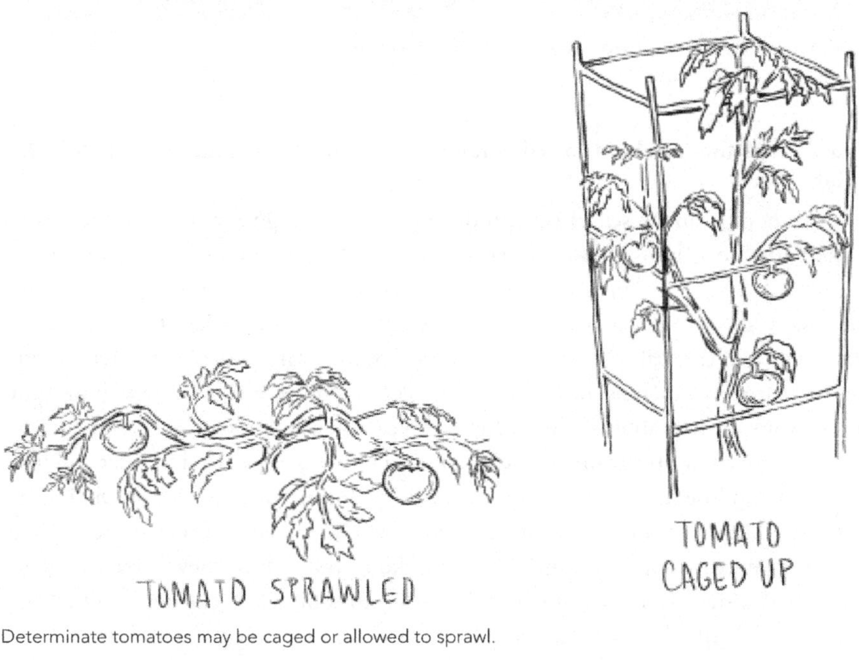

Determinate tomatoes may be caged or allowed to sprawl.

Upright, caged tomatoes provide a better picking experience.

Avoid crops that are hard to tell when ripe and don't continue ripening off the vine

We never want to sell unripe produce that isn't going to taste good, but many pickers just don't know how to harvest properly. Selecting crops that continue to ripen off the vine can provide some insurance against this. Except for the blackberries, every other crop at Pure Land is edible at any size or continues to ripen after picking. It's ideal to post clear instructions and plant varieties that have obvious indicators of ripeness, like a distinctive color change, but nothing will prevent some people from grabbing a firm blue-green cantaloupe and earnestly asking, "Is it ripe??" at checkout. If the melon didn't continue ripening, we would rather cull it at checkout and lose the revenue than allow it to go home with the picker. Obviously, not something we want to do! But it's impossible to supervise every picker all the time. Green cantaloupes still make it to checkout, but at least they'll ripen to good eating with time and we can instruct the picker to wait before cutting into it. If we don't teach the public about growing food, how are they ever going to learn?

Layout

Design the field to be hospitable to visitors.

On a pick-your-own farm, the layout needs to include at minimum a parking lot, a check-in area, the fields, and a check-out area. We'll talk about the other areas more in Chapter 6, but in the actual picking field, aim to create wide middles and pathways. The field should be accessible from multiple directions and have a flow that makes sense.

Keep two feet of walking space between rows at minimum, but three feet is even better to accommodate strollers and guests with mobility aids. For farmers that are used to growing intensively, this feels like wasting a lot of potential growing space, but it's necessary. Non-farmers need a little elbow room to feel at ease in the field. Little kids need space to toddle and plunk down without landing on the plants. Just take care not to overdo it and space the rows too wide, or you'll be leaving money on the table. Also, try not to make the rows too long without a break, or the guests will step on top of them to switch rows and risk damaging the plants.

Keep rows spaced wide enough for strollers.

Plant in blocks.

Most farms likely already do this but try to keep the plants of each crop contiguous to one another. If crops are spread out in a way that doesn't seem to make sense, pickers will miss them. We almost always have bits of row ends here and there left when we run out of transplants, and typically pop a couple spare tomato plants in those, just to use the space. They are never, ever picked from. Keeping crops together also makes giving clear directions easy when someone asks where a particular crop is.

Mulch or lay landscape fabric to minimize mud. Avoid leaving low spots.

Mud and pooled water in the field will stop pickers in their tracks, resulting in unpicked, wasted produce. It's always a relief when the skies clear right before a pick, but that feeling fades fast when there are a hundred flip-flop clad city folk in your sopping wet field. Mulching with straw or wood chips is not only good for the soil but provides some barrier between the guests and mud. In recent years we've started laying landscape fabric between the rows, which was an investment but has worked wonderfully to stop the grass and minimize mud.

Landscape fabric minimizes mud and eliminates the need to weed or mow between rows.

Maintenance

Weed, weed, weed.

It's very important to take care of aesthetics on the farm, including regular mowing, weeding, weed whacking, and tidying. If you are accustomed to leaving tools out all over the farm, make a new habit of putting them away every day. In addition to being unattractive, junk and rock piles are absolutely magnetic to children, so it's best to keep them out of sight for safety. Visiting mothers, of which there will be many, will be grateful.

If the picking field is near anything that attracts snakes, like tall pasture, make sure you're mowing that too. While mulching I once snuggled a copperhead to my chest in an armload of straw; we keep the straw piles well away from the visitors now.

When we used to sell at farmers markets it didn't make sense to spend any time weeding crops like carrots, because I can easily identify their slender, pointy leaves in a patch of nutgrass and henbit and I was the only one picking them. Our guests, on the other hand, have often never seen a carrot in the ground before. Weeding carrots is one of the most time-consuming things I do each spring, but it's necessary if they're going to be picked efficiently. Of course, it's only worth doing because carrots are popular and have a relatively high revenue per foot. They're usually the top answer when we ask kids what their favorite thing was to pick.

Plan to use every weed mitigating tool available. Fair or not, visitors will be

Carrots must be kept relatively weed-free for efficient picking.

comparing your farm to what they see in magazines and on TV. I'm not suggesting you get out the golf course scissors; it just needs to be clear you're making an effort to present a tidy, well-cared for farm and remove any impediments to picking. An online review once noted that our farm was "manicured," which made me and Dad howl. A little mowing can go a long way to impress.

Run cleanup at least once a week

Pickers miss things, which can depress productivity and ultimately lessen yields. The four-pound zucchini that didn't get picked a few days ago is putting a serious damper on that plant's production, so it needs to be yanked and composted (or if still of good quality, donated). This will take a couple hours every week, so be ready for it. This is the entire reason we don't just throw as many plants into the ground as possible; we've developed a sense of how much people will pick, and having to clean up unpicked stuff is just adding more work. It's smarter to determine the volume of crops that can be picked completely, *then* grow more and tweak your crop mix in parallel with the number of pickers to increase revenue.

Be aware that the pick-your-own agritourist does not pick like a farmer. A farmer picks all the cherry tomatoes that are ready on the first plant, and then moves to the next. The agritourist picks one cherry tomato from the first plant, none from the next, four from the seventh, and then moves on. They often don't pick the ripest or biggest, either. This is all part of the nature of pick-your-own, so try not to feel frustrated by it. With sufficient pickers, it all works out.

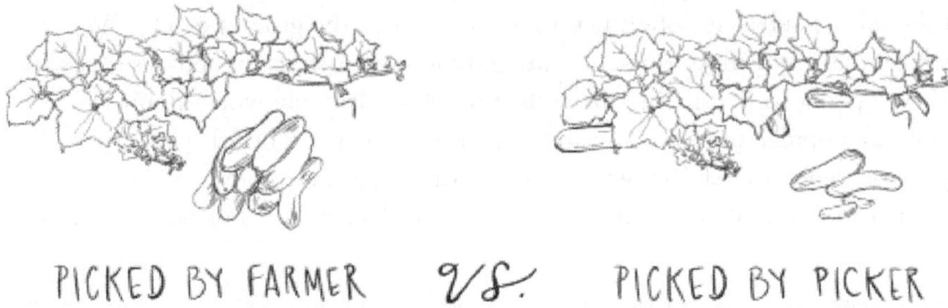

Farmers and pickers do not pick alike.

Have backup seeds and plants ready.

Be prepared for some plant loss due to picker behavior. Unruly or very young children cause some damage every week, yes, but you would not believe how many fully grown humans will yank a plant right out of the ground without knowing what it is. We always start spare transplants so that we can pop another in the hole if it's lost early enough in the season. The same goes for quick-growing direct sown crops like squash and cucumbers, just monitor for damage and pop another seed in as soon as you notice the plant is gone. You've already paid for the row space in amendments and inputs; an empty hole can't make any money! Once it's far enough into the season that a new plant would not be producing before the season ends, we stop backfilling.

It happens.

Ensure all inputs are inconspicuous and safe for use up to the day of harvest

When possible, avoid using products that have residual smells or leave a film on the plants, which can be off-putting to guests. Even OMRI-listed products can have noxious chemical odors that may take a day or two to dissipate, so time your spraying appropriately around picks. If it's unavoidable, use it as a teaching moment; when we spray kaolin clay early in the season to protect young squash, we put a sign at the end of the row that explains what the white powder is.

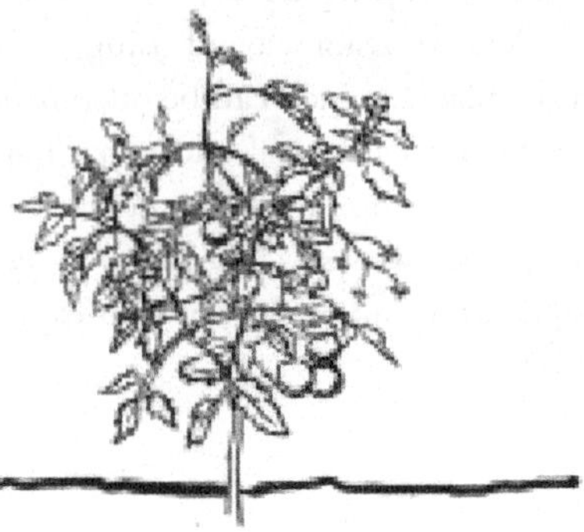

Chapter 5

MARKETING AND FINDING YOUR PEOPLE

The crops are planted and you're *not* signed up to sell at the farmers market. Soon, you're going to have a whole lot of perishable food and you can't secure buyers until it's here - a stressful position for a grower, to be sure. How do you get the word out about your farm to make sure people come and harvest all this stuff? The primary marketing goal for this kind of endeavor is to make sure people know the farm exists and how and when they can participate. In other words, education.

The marketing practices for a pick-your-own are not the same as for any other kind of farm. To patronize our business, we ask *a lot* from people. We're not asking the customer to browse a website at their leisure and click to buy a product that gets delivered to their house. Nor are we asking them to come see us at the farmers market in their neighborhood, where there are lots of other things to do and buy. We're asking customers to get on our website at a specific time to buy tickets, then drive to our place to spend an hour outside essentially *working*, and all they take away at the end are fruits and vegetables (and memories). And we need *thousands* of people willing to do this.

As discussed earlier, different people will have different motivations for visiting. Our product, picking your own vegetables, is one there's not a lot of direct competition for. We don't have to focus a lot on distinguishing ourselves from other businesses that offer the same product because around here there really aren't any; in fact, we often mention other farms to encourage people to support them as well. However, there are a lot of other businesses selling family activities, and even more selling food.

Let's review the motivations and values reported by Pure Land customers from the beginning of the book. Agritourists decide to come to the farm primarily to support local agriculture and make a personal relationship with a farmer, to find the freshest possible produce, to spend time outside in nature, and to do a fun activity.

For some, the product is the carrots. For others, it's the smile or the sunshine.

They value organic, high-quality produce, the fun experience they have, time spent getting fresh air outside, and getting to know us as their farmers. All marketing efforts should tell potential customers that this farm is exactly what they're looking for. *I'm a local farmer, come out to the farm and meet me! I have high-quality, fresh produce! I also have a beautiful place to spend time outside! And it's fun for the whole family!*

If you do your job well, which is to say that you provide a fun, safe experience that offers high-quality produce, the marketing work will largely take care of itself. Word will get around. Your role is to have everything anyone could want to know about the farm readily available in as few clicks as possible. Take for example a mom who saw the farm tagged in a friend's picture of their grinning kids holding up freshly pulled carrots. She came across the post while scrolling her social media and thought, *that looks fun. Pure Land Farm, what is that? *click** Don't let that click take her to a Facebook page that hasn't had a new post in eight months or to an outdated website that still says you're at the farmers market every Saturday. Keep your online presence clear, current, and interesting, and that's all the promotion you will need.

Develop Your Brand

Name

Naming the farm can feel like the hardest part of getting started. A business' name usually makes the very first impression on a consumer, and it's permanent - of course you want to get it right. It took months before we landed on Pure Land Farm, but as soon as we found it, we knew we had arrived; three little syllables that give the consumer a surprising amount of information about us.

Your name should be short, simple, and easily identifiable as a farm. Don't make potential customers work hard to pronounce or understand it. Avoid names that could mislead people into thinking it could be a different kind of business, such as those ending in Hill or Pastures, which could be anything from a company that builds houses to one that makes butter. Generally, including "farm" or "garden" in the name is going to convey the clearest picture of what happens at a pick-your-own. Unless it's a slam dunk, your name is not the place to show everyone how clever or funny you are; you'll have plenty of opportunities for that when you create content. Also, regardless of your growing practices, steer clear of using the word "organic" as the USDA can issue fines for using the word in your marketing if you are not Certified Organic through their program.

Before committing to a name, make sure the domain (or a sensible, intuitive version of it) is available for purchase.

Logo

One Saturday night in our bustling town square I ran into a customer wearing his Pure Land t-shirt. I said, "Hey, nice shirt!" and he replied, "You're the third person to tell me that tonight!" Totally made my day.

Your farm's logo should be memorable, easily recognizable, and attractive. During a conference early on in our farming careers, a marketing professional advised us small farmers and ranchers to incorporate our locations into our logos. The logic is, if all your sales are going to be made locally, make sure everyone purchasing knows *you're local too*. We made the change right away and loved it.

You don't need fancy software to make a logo. There are plenty of free (or cheap) online graphic design applications available for you to take this task into your own hands, and I'm here to say if *I* can do basic graphic design, anyone can. If the concept you have in mind is more complicated than you can handle yourself or needs original artwork, ask around your customer and family network to see if anyone can provide design services for you. In our case, I had the idea for a plant-ey-looking letter P but couldn't create it myself, so a friend recommended a local graphic designer who designed it for a couple hundred bucks, then modified it for us cheaply a few

Adding our name and location to the logo greatly improved it.

years later. If you really don't know where to start and need some inspiration, there are very affordable online crowd-sourcing outfits that will assign your brief to a pool of graphic designers for ideas to get your creative juices flowing.

Adding a tag line can work, but don't make it too long. People shouldn't have to stop and read a logo - it should be fully legible and recognizable at a glance. If incorporating imagery, make sure it's easy to tell what it's supposed to be.

Brand: The perception of a business from the customer's point of view as influenced by all the content the business generates.

People don't connect with businesses; they connect with people. Ask yourself the following questions:

- What do you do?
- Why do you do it?
- What do you want your business to be known for?

Answering these questions is a good way to start defining your brand. Then you can shape the perception of your business by employing a consistent look, feel, and message in all your content. Have you ever seen a post on social media and immediately knew which of your friends wrote it without seeing their name? That's the kind of recognition we want to inspire in our customers with all our communications. Your brand should always speak in a way that makes customers feel like they know you on a personal level.

At Pure Land, we produce high quality fruits and vegetables in a regenerative way and use agritourism to engage with and educate the public about growing food to build support for local agriculture.

Build Your Online Presence

Take every opportunity to provide potential customers with information about you. A dated or bare website and lack of social media makes it much harder for people to find what they're looking for, and they'll give up and move on quickly. It isn't necessary to have an account on every app or to post ten times a day, but making the effort to keep your online presence up to date communicates that you have your act together. No one wants to pile the kids in the car to go somewhere that doesn't appear to have its act together.

The internet is primarily how people will find you and how they will share you with their friends. Some parts of the internet are a one-way conversation, such as a website, where the author is simply projecting information into the ether where someone else may consume it. Other parts of the internet are more like two-way conversations where you can engage back and forth with your guests and potential customers, such as social media. Both are important and serve different purposes.

Prioritize building a strong internet presence over any other kind of marketing. In our experience, in-person marketing events such as vendor shows and health fair booths have never been worth the time and effort they take to participate. At a good event, you may be exposed to a couple hundred guests, engage with a couple dozen, and convert one or two into a customer. The hours spent preparing and attending one of those could have been used to create a lot of social media content, which has the potential to expose your business to thousands of people and has the residual effect of living forever on the internet where people can continue discovering and consuming it.

Word will get out about your farm offline too, where you can't interject with helpful information. We still experience guests dropping by when we're closed because their father's brother's nephew's cousin's former roommate visited the farm and gave them our address. The new visitor never checked the website or social media to learn more, so they didn't know that they needed to sign up for spots before coming. This kind of thing can really slow you down; we've had to stop what we're doing in the middle of the field to greet guests who have wandered in (right past the "closed" sign, before we installed a secure gate), explain how the farm works, load the website on their phones, and escort them back out as kindly as we can. Of course, it's not fun for the guest either, who has come all this way to visit a business and found it closed. Unfortunately, there's no way to reach all of these folks, but we always encourage guests to share how the farm works when they tell people about us. Most will visit the website or social media before jumping into the car, so building a robust internet presence is the most effective marketing strategy.

Website

Building a website is not the soul-crushing task it used to be. Professional-looking websites can be created in minutes nowadays with apps like Squarespace and Wix. Again, I say to you that if I can create and maintain a website, *anyone can.*

Take some time to view tutorials and talk to people you know that maintain websites about the platforms they use before selecting one. If you're not particularly web-savvy, look for intuitive, user-friendly interfaces above the promise of fancy features; most people don't need to be able to change code to fully customize a website. Choose a simple template that speaks to you and allows your content to flow naturally. Look for a picture-driven template that can be loaded up with bright, colorful images of your farm. About half of all website traffic takes place on smart phones nowadays, so ensure that everything works as well on a phone as it does on the computer. To this end it's important to have what's called a "responsive" website, which means it will automatically adjust when pulled up on a smart phone to look nice and function in a user-friendly manner.

Once you've decided on your farm name, the next step is to buy the domain. You can get a cute little .farm domain if you like, but make sure to purchase the .com as well. Most people will assume your website ends in .com, so you'll want to set up the .com to redirect anyone landing there to the .farm.

The first page to create is the homepage, the one visitors will land on when they enter the web address. Your homepage should include the most important information about your business so that a viewer can get a snapshot of what it's all about, even if they click no further. The rest of the content should be populated on subpages and linked with intuitive navigation to the rest of the site.

Your website should, at a minimum, include the following:

- Your story, including who you are and your general growing practices
- Basic information, including your location with a map, a contact email address, and links to your social media pages
- All the activities or services you offer and clear instructions on how to participate
- Produce availability and pricing
- Frequently Asked Questions
- A widget to sign up for your email list

Beyond this, everything else on your website is extra credit. Not long ago, a blog was considered an essential part of a website. Now, the ubiquity of social media has made blogging unnecessary, as it's easier and faster to create content with social media (not to mention much further reaching). But if you enjoy writing long form content like blog posts, go for it. One piece of bonus content we've found useful but not necessary is what we call a "Picking Guide" for each crop, which functions as the place I can send folks when they ask how to store their onions or what my favorite recipe is for beets. Each crop has its own web page with information on the variety, how to pick, clean, and store it, and a few recipes.

Notice I do not recommend embedding a "contact us" form on your website. Include your email address for people to be able to reach you if they really want to, but that's enough; a contact form is a wide-open door for people to holler questions at you instead of looking for the answer on your website. Of course, we don't want them to have to look *hard*. In lieu of a contact form, I recommend a robust FAQ. A comprehensive FAQ will save so much time and headache. It's in the navigation bar of our website but also always linked in the text near the top of the homepage, and I update it frequently, even several years on. Just when I think I've encountered every possible question, a new one pops up. We deal with thousands of visitors and operate in an unusual way; the process of buying from us is not like booking a hotel room or walking into a restaurant for a table, procedures everyone is pretty familiar with. The way we work is not intuitive, so it's our responsibility to share it as clearly as possible.

With an FAQ in place, there is now a destination to send folks with questions. Below is the version of our FAQ as of August 2023:

FAQ

When can we pick stuff?

We are closed until spring of 2024. When we reopen, you must book a spot in order to come pick.

The picking season runs for about 8 weeks from mid-May to mid-July, and then we close until the following May. We pick about 4 days a week. Picks are three hours long and always take place in the morning. They are divided into one-hour time slots so you can choose when you want to arrive.

Stay tuned to our homepage, Facebook, and Instagram, where we will announce the next pick date. We will send weekly emails to keep you informed on what's ready, but do not send emails to announce every pick (sign up for the emails on the homepage in the box that says "Subscribe to our Picking List"). You must check the above sources to know the upcoming open days and times. You cannot book a spot through email or social media messages - all scheduling is handled through the widget on the website.

Here's how it works. We operate on a reservation system, much like a restaurant - we limit the number of people per hour allowed in. We do this to make sure we can keep lines short, our guests have plenty to pick when they come out, and to ensure they can receive personal attention from us.

Before every pick, we will assess the amount of produce that's ready to pick and drop a number of spots in the scheduling widget on our homepage. You can't book a spot more than a few days in advance because we don't know how much produce will be ready more than a few days in advance, and that's how we determine the number of spots.

Admission to the farm has changed! There are two kinds of admission, Regular Admission (single visit passes) and Season Passes.

Regular Admission:

$5 plus tax per person entering the farm one time, whether they are directly participating in the picking or not.

If you plan to visit once or twice during the season, Regular Admission is for you! Regular Admission must be booked in advance on the website for the time slot you wish to visit, you cannot come to the farm without having booked in advance. Spots will become available for booking at a designated time a few days before the pick, the handy countdown timer will tell you when that is. Book one admission for each person in your party, excluding babies under the age of 1. You cannot book groups larger than 6 people.

Season Pass:

We are SOLD OUT of Season Passes for 2023!!

There are two sizes:

$35 plus tax: Individual Season Pass. Admits 1-2 people (this includes children) all season long.

$55 plus tax: Family Season Pass. Admits 1-4 people all season long. Kids older than one year count towards the total of people.

If you plan on coming to the farm frequently, the Season Pass is for you! Season Pass holders are not required to book spots; you can come during any open pick, all season long.

It's not letting me book a spot right now tho!

If you can't book a spot, they have either not been made available yet, they are full, or there are no picks scheduled.

Can't I just wander by one day?

If it's not during a pick, we are closed! Do not drop by, we do not permit walk-ins. When we're closed to visitors, we're working on our plantbabies.

Is there a trick to getting a spot? They sell out so fast!

It's pretty much luck if you refresh the page the moment the spots pop up. The weekend spots go much more quickly than the weekday. One tip is to stay on the page and refresh after 10 or 15 minutes, because the spots will die in people's shopping carts after a certain amount of time if they don't complete the transaction and become available again.

I got a spot, but now I can't come.

Please take a moment to cancel your spots so someone else can use them! All you need to do is reply to your ticket confirmation email saying you need to cancel. To receive a refund, you must cancel your spots before midnight the night before the pick. Please allow a few days for refunds to process.

Your admission fees will not be refunded if you no-show, if you email to cancel your spots later than midnight the night before the pick, or if you bring fewer people than you paid for to the farm. Please understand, regardless of the reason (including illness), we cannot cancel and refund spots on the morning of the pick because there is no time for anyone else to book and use them.

When will you release the spots for [the date I wanna come]?

Spots are only released a couple of days before each pick. You cannot book a spot further in advance, because we cannot predict how much will be ready to pick more than a couple days in advance.

But I want to come in three weeks, and I don't see any sessions listed more than a couple days away. Are you closed? Wha??

All that means is that we have not made the spots available to book yet. We do not release spots more than a couple days in advance of the pick. We have approximately 30-40 picking days every season, so there should be ample opportunity to visit.

That's totally clear and I understand completely! I better get there early to stand in line though, right?

There is no need to get here early! If you've booked a spot, you're getting access to the field at the same time as everyone else. The parking lot does not open until 15 minutes before the first time slot, and we will not start picking before the first time slot begins. Plan to arrive at the beginning of your time slot so you have a whole hour to pick. Last check out is thirty minutes after the last time slot ends and that is when we close, so being on time is especially important if you've booked the last time slot of the day.

So, what will be ready to pick?

It will vary over the season! During May, the thornless blackberries will start ripening along with beets, onions, garlic, and carrots, and into June we ought to have cucumbers, tomatoes, peppers, squash, and more. We anticipate the season closing mid July. The handy chart on the homepage is a good estimate of when crops should be ready. *There is no guarantee of produce availability or quantity from day to day*, particularly for the last time slot of the day. We do our best to target the right number of pickers for the right amount of produce, but we cannot always predict how heavily we'll be picked before you arrive.

Sweet. How much does it cost?

Depends on the crop! For every picker there is the $5 admission fee (counted as feet on the ground - there is no fee for babies). Blackberries are $7 per pound, big tomatoes and beets are $3 per pound, cherry tomatoes, carrots, and sweet peppers are $4-5 per pound, hot peppers are $8 per pound (a pound of hot peppers is a lot of hot peppers), onions are $2 per pound, cucumbers are $2.50 per pound, garlic are $1-2 each, squash are $2 per pound, sunflowers are $1-2 each. Our prices are aligned with the other local growers at our Dallas-Fort Worth Farmers Markets. You may pick as much or as little as you wish, but if you pick it, you bought it, so keep an eye on little ones and don't pick more than you can use. If we see you picking and ditching good produce in the field, you will not be permitted to return.

How does it work?

When you arrive, you'll line up to enter and check in by name. You'll get a bucket or basket from us under the tent, and we'll talk about what's good for the pickin' that day and how to do it. Then you can head into the rows to make magical memories harvesting beautiful fresh produce with your family. There will be signs and help in the rows to tell you what's what and how to pick it. We've even set up QR codes this year you can scan to show you a video of how to pick.

When you're done, come on back up to the tent, tag us in a sweet selfie with your haul of veggies, and we'll weigh the crops out to total you up.

Do I have to bring cash, or do you take credit cards?

It's 2023, y'all! You bet we take credit cards! But they do take a chunk for processing fees, so cash is always king.

Awesome! What should I bring and wear?

A cloth or washable bag would work great to take your veggies home, but we do hoard paper bags from Trader Joe's and Whole Foods all winter too. If you have a bag pile too, we'd love you to bring it in for us to reuse!

A hat and closed-toed shoes are ideal attire. We wear long sleeves and pants even in the summer, so bring sunscreen if you need it. If it rained the day before, boots are a good idea. While we do not generally have chiggers, bug spray is also always a good idea.

OMG, will there be bugs??

Totes! We do NOT use synthetic pesticides, so you are definitely gonna see a few bugs, many of which (whom?) are actually beneficial!

Can I bring my dog?

Sorry but for food safety, no doggos are allowed at the farm.

What happens if I need to pee?

We have two wholly adequate Port-a-Potties that are serviced weekly.

I can handle that. But what if it rains?

We will stay open for sprinkles, but not for thunder and lightning. If the weather is looking bad, we'll post a cancellation on social media and you'll receive an email and text canceling the pick. We will not make the decision to cancel any earlier than the morning of the pick. You just can't trust the weatherman!

Do you grow or sell [x]?

We do not grow blueberries, corn, pumpkins, and many other things. We do not have chickens for eggs or cows for milk. Currently, we only grow the following:

blackberries
carrots
beets
onions
garlic
cucumbers
bell peppers
hot peppers
slicing tomatoes
cherry tomatoes
cantaloupes
summer squash
sunflowers

Do you sell direct to the public? Can I place an order and pick it up? Do you wanna join my farmers market or supply my farm stand? Do you sell to restaurants?

We are a one hundred percent you-pick farm, meaning we only sell directly to the public in a fun agritourism setting. You book a spot in the time slots we offer as described, then you come out to the farm to pick and pay for it. You cannot place an order to be picked by us, and then picked up or delivered. We do not go to farmers markets or supply produce in bulk. We do not keep a farm stand or sell to restaurants. *Youuuuuuu*-pick!

Hey, why didn't you answer my email or Instagram DM or Facebook message or TikTok comment or phone call or voicemail?

We are a two-person company here at Pure Land Farm! That means there is no "office person" manning the phones. We keep pretty busy planting, mowing, trellising, fertilizing, weeding, pruning, mulching, shoveling, spreading hay, inspecting crops, applying pest and disease control, scheduling picks, updating the website and social media, and washing baskets.

Search Engine Optimization (SEO) and Google My Business (GMB)

There's no point in spending a lot of effort to build a website if it doesn't come up on an internet search. If someone tells their friend about my farm and the friend later searches for it by name, Pure Land Farm ought to be right at the top of the results. But if the friend forgot the name, they might search something like "pick your own farm McKinney." To make sure Pure Land comes up in that case, we use Search Engine Optimization (SEO). Whole careers are devoted to SEO work; we are only going to scratch the surface here with the simplest ways to tap into it.

One way that SEO works is by identifying "keywords" on a website and their frequency of use. The keywords are entered by a user in the search field, and then the search engine scans the internet for websites where those words appear most frequently, to suggest them as the highest or "most relevant" results. What keywords might someone enter to find your farm? Include them often in the text and headers on your website. Think pick-your-own, farm, vegetables, organic, local, family, fun, and the like.

As the premier search engine, Google is where most people will turn to seek information about your farm. Google provides a platform for businesses to control their information called Google My Business (GMB). Setting up a profile is simple and accessed through the GMB portal after signing up for a free account. There, you can add business details like hours and location, post updates and photos, respond to reviews, etc. It's important to set up a GMB page because in response to queries, Google will present the information in what they call a "knowledge panel" aimed at providing a summary to users at a glance. GMB is also how your location will be pushed to Google Maps, the most popular navigation map in the US, which many guests will likely be using to get to the farm.

Merch

Invest in some promotional t-shirts, hats, stickers, reusable tote bags, and the like, but don't expect to profit from them. Merch should be considered more of a marketing expense than a source of income. Think about the people in your community you'd like to see walking around in your t-shirt, ask them their size, and give them away. This is cheap, hyper locally targeted advertising. Everybody likes a free t-shirt! Just make sure they're soft and comfortable—the most economical options can be pretty rough. Nobody likes a scratchy t-shirt.

Depending on your available bandwidth, the most profitable option is to order these items in bulk, set up ecommerce to process sales through your website, and fulfill orders yourself. This is time consuming, requires a big upfront investment in the stock, and may leave you sitting on a pile of merch for a long time. We prefer the newer print-to-order options where you can set up an online store with a third party vendor that handles all of the payment processing, printing, and shipping on their own. It's expensive - in fact, we make no profit at all, choosing to keep the price lower to encourage sales rather than pocket a couple of dollars per shirt - but it's worth it to us for the sales and fulfillment to be completely hands-off.

Social Media

Social media allows us to consistently remind our fan base that we exist and we're awesome. According to Statista, the average American spent two hours and three minutes on social media every single day in 2022. The popularity of individual apps will wax and wane, but whatever the dominant social media platform of the day, you have to participate; in 2023, commit to Facebook and Instagram as a baseline. Set up social media accounts with your logo and basic information and ensure the profiles link back to your website, which is ultimately where you want to drive people.

When you're just starting out, begin building a social media presence well in advance. It is not advised to sign up for Facebook as your first 400 pounds of cherry tomatoes are blushing; you're going to want a sizable following by that point. Use the cold and rainy days well before your season begins to create a digital presence and just start posting.

Content: Any piece of media created or shared on a website or social platforms, including images, text, links, videos, etc.

Voice: The style and language your business uses consistently across all forms of communication. Think of it as your brand's personality.

Creating Content

Now for what many people consider the hardest part. What are you supposed to *say?* Quality content creation is tough, and it's easy to feel like there's nothing to post when you're knee-deep in the unglamorous daily chores of farm life.

Your content should all sound like it's coming from the same character, your farm personified. Essentially, this is your brand's voice. The best advice for honing your brand's voice is to stay on topic and be yourself. Like an eighth grader with a British accent after a week abroad, you're not going to be able to keep up a facade for long, so don't try. Talk how you actually talk, and talk about things you actually care about. *Authenticity is interesting.* If at all possible, be entertaining and funny as well; at least, your funniest self. Leave anything unrelated to the business for your personal pages. Needless to say, steer completely clear of politics and divisive, hot-button issues.

In deciding what to post, first consider the nature of people willing to spend their hard-earned money and free time going to a farm to pick vegetables. These are people who love local agriculture, food, spending time with their families, and getting outside. Remember, most people scrolling social media are tucked behind desks all day trying to avoid work. There's a reason why travel influencers are so popular on social media; they provide *escape*. Farms do too!

Content pillars are a few topics relevant to your business that represent the core of what you do. By creating a framework like this, it will always be easy to find something to post. For example:

Within these content pillars, fill in specific post ideas based on what you're up

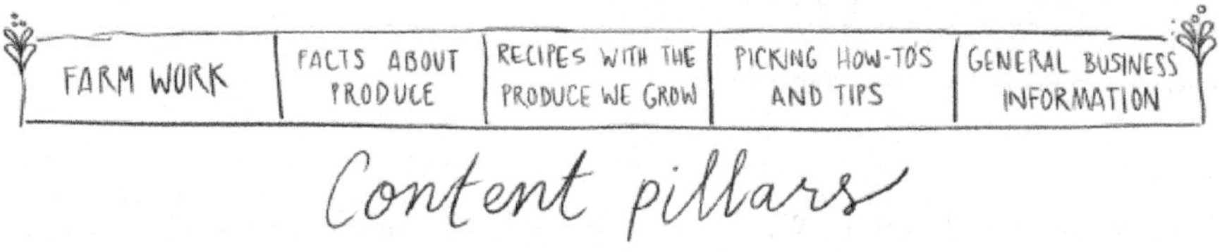

Some sample content pillars for our business.

to that week. I keep a running list of post ideas in my phone, capturing them when they strike me to save for later.

Then, select and shuffle posts from each content pillar to create a steady stream

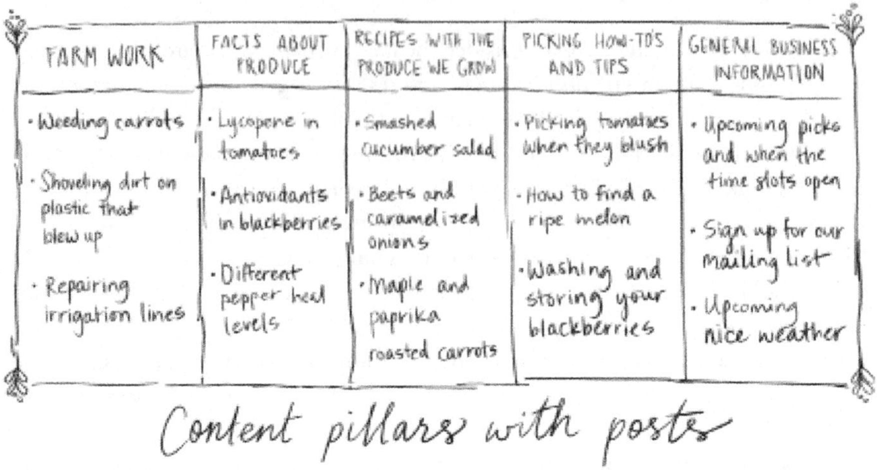

Ideas for posts in each pillar.

of interesting content for the week.

Some people find it easiest to post spontaneously, and others prefer to plan

A full week of relevant, interesting content.

their content ahead of time. We do a little bit of both. For example, I might schedule a recipe post in advance, as I'll take my time writing the directions carefully and snap a picture during the cooking process. But people really enjoy spontaneous "slice of farm life" posts too, where we stop for a minute during our daily work and share what we're doing, so I try to do that every day.

Images and videos will typically get much more engagement from your followers than posts that are just text. Here are a few tips for creating great photo and video content.

Photos

Entire apps have been built around the human desire to look at beautiful things. Luckily, farms are full of beautiful things, so finding subjects worthy of photographing is the easy part. The slightly harder part is capturing a subject and turning it into an attractive, compelling image. With a little practice, it will become easy to turn a just-okay photo into a great one.

Quality

Incorporate shooting pictures and videos into your daily routines.

- It may seem counterintuitive, but bright sunshine does not make for high-quality photos. On bright but cloudy days, take a few extra pictures around the farm to bank some content.
- To focus on small things, like a budding leaf or a little bug, place your palm behind it, allow the phone to focus, then pull your hand away and quickly snap the shot.
- Always take an extra minute to adjust the levels in your pictures in your phone's photo app, a third-party app, or in the social media posting process

itself. You would be surprised how much a photo can be improved by sliding a couple of bars to modify the warmth, contrast, etc.

Composition

- Typically, a photograph contains a subject and a background, which should be arranged in the frame to form a pleasing image. Make sure it is obvious what the subject is by being reasonably close to it and by placing it appropriately in the frame for the destination image's aspect ratio, which is the width of an image (or video) in relation to its height. For landscape photos, consider using the rule of thirds, in which the subject is placed one third of the way into the frame from the left or right, usually resulting in a pleasing composition. For squares and portrait orientation, centered subjects generally look good. Use your intuition to seek balance between the placement of the subject and the background.
- Annoyingly, aspect ratios come in and out of fashion with different apps and

A centered close-up of some beautiful berries makes a compelling image.

features, so capture multiple shots with different aspect ratios to cover your bases. Instagram is famous for the square, which has an aspect ratio of 1:1. But Instagram Stories, a feature launched much later, have an aspect ratio of 9:16. This is also known as portrait, and it means a typical square-shaped Instagram image will not fill an Instagram Stories screen. It's not the end of the world, but it can distort or crop some content to the point of making it unusable. Only recently have I gotten into the habit of taking the same picture in both landscape and portrait orientations to use the appropriate version where it fits best.

- Don't forget about the background! Avoid shooting subjects in front of backgrounds that are busy or messy. A photo of a juicy red tomato on the vine can be ruined by a thatch of dead brown leaves behind it; sweep them out before snapping the picture.
- People like pictures of people, so get out from behind the phone or camera occasionally to include yourself and your farm team. Sorry, I don't make the rules.

Moment

- Always ask yourself what you want the picture to say before you decide how you're going to shoot it. For example, if you want to say that there are lots of shishito peppers ready to pick, the image should show a particularly loaded plant where the peppers are clearly visible, as opposed to a shot taken ten steps back that captures the whole row but where you can't see the peppers.
- Keep it light, fun, and aspirational. Share the tough moments too but keep

Decide what you want your photo to say before you shoot it.

them rare.

Videos

Videos can feel paralyzing, especially if you're not a natural entertainer or used to being in front of a camera. Unfortunately for most of us farmers, videos are digital gold to social media apps; they earn preferential placement from the algorithm to get wider visibility in your follower's feeds. In other words, an image might be shown to a couple hundred people, but a video might be shown to a couple thousand. For that reason alone, they're worth pursuing. Consider all the advice on images in addition to the following when it comes to shooting videos.

- Center yourself fairly close to the camera with your face two to three feet away.
- If you're uncomfortable having someone film you, figure out how to do it on your own. I, personally, would rather die than suffer the embarrassment of having Dad film my dumb social media content, so I have a couple of tripods. They're cheap and come in all kinds of configurations.
- Practice first, but don't practice so much that it feels canned. Run through what you want to say out loud once or twice and then roll the camera.
- Keep it short and get to the point *very* quickly. Marketing studies have shown the average attention span for consuming this kind of content is well under 10 seconds before the viewer swipes on; try to pique your audience's interest in the first line or two so they'll stick around. Be sure to remove any dead air at the beginning before you start speaking as you will lose people immediately.
- Be your authentic self, don't try to project how you *think* you should appear. Remember, the audience wants to get to know *you.*
- Don't let perfect be the enemy of good. Just do your best; practice it, shoot it, post it, and move on. No one expects you to be a polished social media star. Some editing is necessary,

Don't be scared of posting videos. Find something interesting to talk about and just go for it.

but it's easy to lose half a day editing a video which may not become any more impactful to the viewer.

Reuse the same content on as many platforms as you can, known as "cross-posting." For example, some apps like Instagram can be set up to automatically push the same post to your Facebook page. Content from individual apps can also be published to your website through a block or widget, which will provide a steady stream of fresh material to your website with no additional work. Some great content just falls into your lap with no work at all, like when you are tagged in a visitor's cute pictures from the farm, and all you have to do is hit the "share" button.

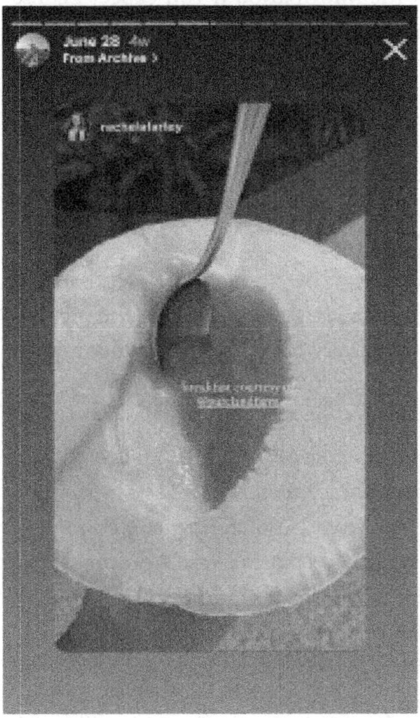

Share or repost social media content from your followers.

A generally attainable target is to post once daily, increasing to two or three times a day during the picking season. Find the sweet spot between committing to a posting schedule you can't keep up with (which will stress you out) and one that's all over the place where you go a week without posting anything (which will also stress you out.) During the off season, we go months without posting, then rev the social media machine back up in early spring. It turns out, the more we post in the fall and winter, the more people ask if we're open, and it doesn't feel or look great to answer that you're currently closed over and over. Someone just happening upon the page might assume you're *never* open, or see a recent post and say "Hey look, a farm!" and jump in the car to find us totally closed, also a problem.

Managing Inboxes and Comments

Because we engage with so many people and our business entails as much hospitality as farming, we get far more messages, questions, and comments than the average market garden. I always smile during picks when a guest introduces themselves, saying, "I'm the person that messaged you yesterday!" *You and 40 others, friend.*

Free yourself from feeling responsible for answering every single message you receive. There is an expectation that like lots of other businesses, there's someone at the farm doing general administrative work like answering the phones and replying to messages. This is not the case at our farm and likely won't be at yours either. Bigger, seasonal agritainment farms may have an "office person" filling this role, but we run pretty lean and keep busy growing as much food as we can. For this reason, we discontinued phone service to the farm a few years ago. Leaving a voicemail reasonably leads the caller to expect a call back, but answers to the vast majority of inquiries we receive are easily found on our website. Returning ten voicemails asking for directions to the farm is simply not a good use of time.

However, it *is* your responsibility to provide information if you expect people to do business with you, so put systems in place to answer as many questions as possible *before* someone feels compelled to send a message. The more information you provide in more places, the greater the possibility someone will stumble upon what they want to know before contacting you. When there's a social media comment asking a common question answered in the FAQ, respond with a link to it. Even if the asker doesn't click to read it, someone else scrolling by might, and every set of eyeballs that reads even a snippet of the FAQ is a win.

Most social media apps and email providers can be configured to send an auto-reply when a message is received. *This is your new best friend.* During the picking season, this is our auto-reply message. Can you guess what the most frequent questions are?

> Hey, thanks for contacting us! This is an auto reply, we are unable to answer messages during picks. If we don't reply to this message, your question is answered in our FAQ: https://www.purelandorganic.com/faq. If you're wondering how to book a spot, it will tell you. You must book a spot before coming to the farm. If you're messaging to try to book spots because they're sold out, we're sorry - they're sold out!

During the off season, this is our auto reply message:

> Hey, thanks for contacting us! We are CLOSED for the 2023 season, reopening in May of 2024. If we don't reply to this message, your question is answered in our FAQ: https://www.purelandorganic.com/faq

---

With these auto-replies in place, if someone sends a message that's answered in the FAQ, we don't reply. This cuts down on about ninety-five percent of the messages we need to personally answer.

Remember that your social media is *your* turf, so delete or ignore any public comments with impunity. It isn't necessary to argue with unpleasant commenters or reply to the truly baffling people that will simply comment the word "address" on a post instead of clicking on the page name to see the address. Keep an eye out for spam comments and promptly delete and block them; spam left posted publicly on your accounts makes it look like you're not in control. Social media can feel like a thousand people screaming for your attention, but don't get bogged down. Delete, ignore, block, and move on.

Managing Reviews

Google, Yelp, and some social media apps allow people to leave reviews for your business. This is nerve-wracking, of course, but I'm always suspicious of businesses that have the review function disabled. When starting out, ask friends and family to leave the farm some glowing reviews to get the ball rolling.

Managing good reviews is easy, a genuine pleasure. On rough days, I'll scroll through our five-star reviews for a little pick me up and to remind myself why we do this. Always "like" the good reviews and leave a short reply (if you remember the guest, a personal touch makes it even better, like "I hope the mud came out of your daughter's dress"). During picks, when a guest has clearly had a good time, get in the habit of asking them to leave you a review as they leave.

The bad reviews, well, that's a stickier wicket. I always feel it's best to reply in public to bad reviews rather than ignore them, because if you don't, anyone who reads it only has one side of the story. But never, ever lose your cool. Thank the guest for coming, tell your side of the story succinctly and without blaming the guest (this is, of course, for mildly negative reviews - I've yet to have a bad review accuse us of anything worse than being picked out of something they wanted or the spots filling up too fast). Apologize for the disappointing experience and let them know you'd love to have them back for another try.

You cannot please everyone, and it's unrealistic to think otherwise. Even the finest restaurants in the world receive 1-star reviews from the unaccountable yet highly opinionated internet commentariat; do not tie your self-worth to a couple of bad reviews.

Email Marketing

Email marketing is an excellent way to directly reach your customers and one you're probably familiar with if your inbox looks anything like mine. There are lots of email marketing companies to choose from, such as Mailchimp and Constant Contact, and the cost is typically based on the number of email addresses in the database. Select a system that allows folks to "subscribe" to your email list through a widget on your website, one that's seamlessly integrated so the sign-ups are automatically compiled without any intervention from you. Position the sign-up widget in a highly visible place on the homepage where it may be seen without scrolling, not near the bottom where it might be missed. Start collecting email addresses early, as soon as the website is launched.

Taking an hour to set up an email template will save lots of time over the season. A template is just a standard email format with your logo, links, and content blocks all in place. Then you won't have to build an email from scratch every time you want to send one, you can just select the template and change out the relevant information.

During the picking season, we send weekly emails to announce upcoming pick dates and what will be available to harvest. We also always include a personal note that talks about what's going on at the farm and links to the website and FAQ. Never send an email without giving the recipient something to click on to take them back to the website, and if it's during the picking season, always include a clear "BOOK" button that leads directly to the booking page.

Due to our relatively high demand, we have to be careful with the timing and frequency of marketing emails. We never send them right before the weekend because we know spots for the weekend always sell out and we don't want more unhappy customers that couldn't get spots. Instead, we send them at the beginning of the week, when we need the awareness boost to fill the weekday picks. Sending too many emails will lead to unsubscribes, so we generally stick with one per week as long as all the picks fill up, adding a second if necessary.

Leverage Your Community

While it's never a bad thing to have a big online following, it's more valuable to have a small one geographically close to you than a big one spread out over the world. Only those nearby can actually come to your farm and spend money! So don't obsess about trying to go viral or get on Oprah (unless you're farming in Montecito.) Instead, devote most of your energy to developing relationships with the people right where you are.

When you open the gates to let the public into your farm, it becomes part of the surrounding community, and therein lies great power. When you're part of a community, *your* voice is not the only one that can speak for you; your story can be told over and over through the voices of others in the community as well, from one individual to another all the way up to institutions with the power and influence to reach thousands. This is true, meaningful *reach*.

Individuals

Your customers - in particular, your regulars - will be your biggest cheerleaders. I can't tell you how many times a person has come out for the first time to pick on their own, then brought a friend the next time. You won't have to ask your regulars to spread the word about you, they'll already be doing it. Share their posts, reply to their comments, and make them feel like the vital members of your farm's family that they are.

In 2020, as the COVID-19 pandemic closed schools and businesses, everyone became desperate for outdoor activities and our popularity shot through the roof.

One mom stopped me during a pick and said, "I don't know if you know this, but your farm is THE activity everybody wants to do in my Facebook mom group. When people get a spot, they brag about it. I can't wait to post that I was actually here!" Dad and I were floored. With absolutely no effort from us, this group of over 10,000 local moms was taking care of all the advertising we needed just by posting about us to each other. You don't need to wait to be discovered, you can proactively join these groups and post about your farm. Like death and taxes, it is certain that moms will be looking for activities for their children until the end of time.

Local media

Small, local online or print newspapers and periodicals are always, *always* looking for content, and pick-your-own farms are the kind of "feel good" story people love to read about. Dig around for the email addresses of the reporters and editors at these outlets and pitch them to do a story.

There are a couple of things to look out for when working with people who say they want to write about you. Many publications rely on ad placement for their revenue, but don't get talked into buying an advertisement; you don't need it. *You're the story.* Ask if you can see the final piece before it's published to ensure the details are correct and the spin is positive. Also, take care to only spend your precious time talking to writers who will reach your target audience. Spending an hour with someone from a free local paper writing an article about fun summer activities in your area, yes! Spending an hour with a college student writing a blog about farming for a class project, probably not.

One non-traditional media channel with the potential to reach a lot of people is the growing pool of social media influencers. It's easy to dismiss influencers as kind of absurd, but "influencing" has become a legitimate career in the last few years. Essentially, influencers are considered authorities by their audience, who become "influenced" by them to make purchasing decisions. The influencer projects aspiration: *you can look as gorgeous as I do by using this product, and your life can be as fabulous as my life by going to these places!* Specifically, seek out the food, lifestyle, and mom influencers in your area by poking around apps like Instagram and TikTok. Any time an influencer with a good-sized following reaches out, invite them to the farm to partake in the picking and make sure they have a good time. No, you don't have to give them anything for free. If the influencer has fewer followers than you do, you can always pass.

Businesses

We partner with a couple of small businesses that sell complementary items we do not, like baked goods, during our picks. This is a great opportunity to cross-promote to each other's databases, which exposes one another's business to a new group of potential customers.

Pickers will often ask for a recommendation on where to have lunch after their

Pop-ups by local bakers can be a popular attraction.

visit. If you continue to do sales to any restaurants or know any genuine farm-to-table restaurants in your area, send them there. Supporting restaurants that buy directly from farmers benefits both the local farming community *and* a restaurant that's doing the right thing by sourcing locally.

Other Farms

As you convert the tourists visiting your farm into regulars and local food true believers, they'll start to ask you where they can find farms producing meat, eggs, milk, and more. Find local producers with similar standards to yours so you can point people in their direction with confidence. They will point people *your* direction, too. Supporting fellow regenerative farmers should always be a priority, as doing so strengthens the local agriculture economy for everyone.

Nonprofits

When you can, give back to the nonprofits serving your community. We often donate Field Trips as an auction item for fundraisers, which only cost us an hour

A post-season glean of the field to donate to local nonprofit farm McKinney Roots.

Food pantries are often in need of fresh produce.

of time but can raise a couple hundred dollars for the nonprofit and promotes our business to their audience as a bonus.

At the end of every season, we host a "donation pick" where we enlist volunteers to pick out all the remaining quality produce from the field to give to nonprofits feeding the hungry. It's obviously the right thing to do with the unsold product but has the added benefit of yielding tremendous goodwill from your customers and the community. You can also call these organizations during the season if the field gets ahead of the pickers and you need to glean off big squash or have too many ripe cantaloupes at once - they may even send volunteers to pick it.

Chapter 6

RUNNING YOUR PICKING SEASON

After months of cultivating the crops and building an audience, your tomatoes have started to blush and there's a line of eager pickers at the gate. *It's happening.* This is the fun part, but it's not as simple as throwing the farm gates open and yelling, "Come on in!"

> Thy small farm shalt:
>
> BE A PICK-YOUR-OWN OPERATION.

To be appealing and enjoyable to visitors, it's important to present a well-planned experience from beginning to end. Having noticed the public's enthusiasm when we started doing pick-your-own, a couple of nearby farmer friends did a little dabbling themselves. One set out an unmanned stack of baskets, a scale, and an honor cash box then headed across the farm to work, leaving any customers that might stop by to fend for themselves. Another posted a one-hour block of time and invited people to drop by to pick from just *one* crop. Neither farm had many takers. Remember, this endeavor is as much tourism as it is agriculture.

Agritourists are looking for an organized, friendly, fun experience. For many, yours will be the first farm they've ever been to, and they probably won't have any idea what to do or expect - not just the children, the adults as well! These customers are bringing their families out to see you; they want it to be worth their time, effort, and money. Try to put yourself in their shoes and play out every aspect of what the experience will be like from their perspective. Is the booking process as smooth as it can be? Where will they go first when they arrive at the farm? Is that obvious, or does it need another sign? Can they find the restrooms easily? Then, once the season is open, be on the lookout for any kinks to eliminate to make the whole operation run more smoothly. It's probably not going to be perfect on the first try (although, with this book, I hope it's closer to perfect for you than our first season was).

There are two main goals during the picking season:

- Everything gets picked. Some waste and loss is inevitable, but the majority of the crop yield needs to be harvested by pickers to meet the income potential.
- Everyone has fun. When guests have a good time, they come back, and they tell their friends.

At this point, you've put a lot of effort into growing high-quality produce; now you need to gather a few materials, think through some tough decisions, and put a little effort into hospitality. Fasten your seatbelt; you're about to go for a wild ride.

Ready

What infrastructure and equipment are required to operate a pick-your-own farm? As it turns out, not a lot.

At Pure Land, we have no building structures to speak of besides an old barn where we store implements, a shed full of tools, and a travel trailer with luxurious amenities like a mini fridge and microwave. We used to feel a little self-conscious about these relatively austere conditions. *It's not much.* In our minds, visitors to a "farm" would expect a big homestead with sweeping wraparound porch, picturesque red and white barn, the whole enchilada. But we were working with a chips and salsa budget; the investment in farming equipment and materials alone stretched us too thin to even think about building any permanent structures.

Now, we wear our frugality with pride. It simply does not take a building of any kind to do what we do. There's no need for a commercial kitchen, pack and wash station, or walk-in cooler. Sure, it might be nice to have visitors pull up to a pretty barndominium, and if you have one I'm still jealous, but it's totally unnecessary to go into debt that way. We could pick up and move the whole thing to a new location any time, so it isn't necessary to own the land either. For the business we are currently running, spending money on costly infrastructure like a building would not increase our revenue, so we'd rather keep it in our pockets. It appears that the customers don't miss it, either - being inside is not what they've come for.

My favorite thing about pick-your-own farming not requiring expensive infrastructure is that it removes one of the biggest barriers to entry - a large amount of startup capital. *Anyone can play.* You need some dirt, the equipment necessary to grow food plus a few extra bits and bobs for the pickers, a laptop with an internet connection, and depending on your size, a little help. Some soil, some stuff, and some staff.

Stuff

From here on, it's assumed you have the land and necessary farm equipment squared away. On top of that, these are the tangible things you need.

Tents

Here in North Texas, there is never enough shade. We set up large metal pole tents to cover the check-in and checkout areas. Pickers spend less time in the field on hot days, often picking less than they intended to, so choose tents large enough to provide room for folks to cool off and hopefully head back out to do a little more picking. Make sure to size the checkout tent large enough to accommodate a line and order at least one extra tent top for each tent, ready to deploy when they become damaged.

Tents provide shade and tell customers where to go.

Tables

If you're already a market gardener, you probably have tables from your farmers market booth. We're partial to the bifold plastic variety with a handle on the side for easy storage and moving around - in fact, we're still using the same tables we bought over ten years ago when we started farming. Whatever kind you choose, make sure they can easily be hosed off and scrubbed, because they will get dirty. At our size, we use about eight tables between check-in, checkout, a spot in the shade to hold full baskets while guests go back out to pick more, and spot for guests to organize their baskets as they approach checkout.

Fans

We have a couple of big ranch fans inside the check-in and checkout tents, which may not be necessary if your climate is more hospitable than ours.

Baskets

Regular two-handled grocery store shopping baskets work great for pickers to collect their harvest. The slats allow loose dirt to fall out, but make sure they're not so wide that a cherry tomato could slip through. Any baskets you provide should be easy to clean and stack. Don't be cheap with the baskets; buy more than you think you need. Kids often want to carry their own, and we have many customers that will fill one basket, set it down in the shade under the tent, then grab an empty one to keep picking. *Enable this behavior as much as you can.* If you're going to grow something delicate like blackberries, you'll need to provide a separate basket for them to avoid crushing. You'll often see these referred to as "pails" by online retailers.

Provide baskets for pickers to harvest into.

Picking tools

The more tools you put out for pickers to use, the more you are going to find later in the field with the tiller. If a tool isn't needed for picking, don't provide it. Scissors are the only tool we've found necessary for the range of crops we grow. If you grow something that requires digging you might consider small spades as well; we keep a bucket of them by the carrots. Many guests like to bring their own gloves, tools, and even baskets, and that's fine.

First Aid Kit

A basic first aid kit containing bandages, disinfecting pads, and topical pain relief creams comes in handy for scrapes and stings.

Bags

Farmers market shoppers have been well-trained to bring their reusable totes to market, and you should encourage your visitors to do this as well. But many will forget or pick more than the bag they brought can hold (hooray!). The best bag we've found for produce is the typical two-handled paper grocery sack found at most megamarts, one with stiff sides that can stand upright on its own. Several seasons in, we've actually managed to train our customers to save their own paper grocery sacks during the winter and bring them to us when the season opens, so we haven't had to buy bags for the last two years. We're saving money and recycling! Of note, we do not provide plastic clamshells for our blackberries. We started off that way but found out quickly they are hella expensive, and it turns out blackberries travel home just fine in bags as long as they're not overfilled. We line the blackberry pails with a thin plastic or compostable bag so guests can pick directly into it, then we can simply lift the bag out onto the scale to weigh and tie to send home, minimizing handling of the delicate fruit.

Bags saved by pickers over the winter are brought to the farm the following season.

Use shallow tubs to make it easier to weigh produce.

Legal-for-trade scales

To sell produce by weight, it's required to use a scale that is certified as "legal for trade." This means it has been tested for accuracy to assure the seller is charging appropriately for the product. Some states require sellers to obtain a certificate from a technician that validates the scale is measuring accurately (Texas is not one of them). Look for scales that can measure from less than a tenth of a pound up to fifty pounds and are battery-operated to give the most flexibility in placement. Tabletop scales usually have a pretty dinky platter, the flat part on top where you set whatever it is you're weighing. What works best is to find a wide, shallow tub to place on top of the platter, then pile unwieldy vegetables into that. Just make sure to zero out or "tare" the scale with the tub in place to remove its weight from the total. To keep lines down, we run with two or three full checkouts every pick, complete with their own scales, credit card readers, and cash boxes.

Credit card readers and cash boxes

About two thirds of our guests pay by credit card. The fees are brutal, but they're a fact of life if you're going to accept payment by card. There are several options for reader hardware, so I recommend using the one that is best integrated with your existing accounting systems. We use the Quickbooks mobile payments device, a small, inexpensive, rechargeable device that accepts card payment by tap, chip, Google Pay, and Apple Pay. We have thus far avoided having to accept payment by digital wallet apps like Venmo and Cash App, but I recommend setting accounts up just in case the guest has no other way to pay or forgets their wallet (it has happened more than once). Prepare a standard cash box full of change for every pick as well: $100 broken down into 40 one dollar bills, six five dollar bills, and three ten dollar bills, plus a couple rolls each of the coins.

Bathrooms

It takes two standard portable restrooms to accommodate the number of visitors we have every week (about a thousand) without getting too funky. Make sure the rental company will service them weekly outside of the hours when pickers are on site - nothing stinks more than a port-a-potty being serviced, literally. If you can't provide access to a sink with potable water, handwashing stations are usually also available to rent from the potty provider.

Signs

Every sign is a chance to reach someone with information. Think of a sign as the opportunity to not have to say something out loud. It doesn't sound like much, but after telling the twentieth person in a row that the check-in tent is *over there*, you'll see what I mean. In addition to the Texas Agritourism Act warning signs we are required to post, we have signs all over the farm. *There can never be enough signs.*

The first sign everyone will see should be at the entrance from the public road, a large one with the farm name and logo that tells visitors they're in the right place. Check with your local department of transportation to see if you're required to keep the sign set back a certain distance onto your property from the road. If your farm is off the beaten path, see if you're permitted to put a small wire stake lawn sign on the nearest main road with helpful directions.

We use a large sign to indicate where the check-in area is, one that can be seen at a distance from the parking lot as guests get out of their cars. Inside the check-in tent, we have a sign that explains how the picking experience works and one displaying the farm rules. We have another sign with the prices posted where guests pick up their baskets to ensure they are informed of the costs before they start picking.

On each row, there is a small sign designating what is planted there and brief, clear picking instructions. Depending on the field layout, you may need signs on both ends of the rows. It's helpful to have a few "do not pick" signs ready to deploy quickly in the field when needed, too. We've taken to printing these on red paper to draw attention. These would be put out when the melons have hit size but aren't yet ripe, or when the squash have been picked down to nubs and need time to recover.

Signs in the rows tell pickers what is planted and how to pick it.

We use a variety of materials to make signs, mostly because we are cheap and like to use what's laying around. The large entrance and check-in signs are professionally printed on vinyl but the rest are squares of plywood nailed to wooden stakes, metal real estate signs we've painted over, and laminated sheets of paper taped to garden markers. The latter I've found especially useful in the field as they can hold a lot of information including QR codes or a photograph of exactly what the picker should be looking for, like the partially opened stage at which you should pick a sunflower.

Use what you have to make signs.

Online Stuff

In addition to the website and social media you've already set up to start marketing, you'll need a couple of apps:

Payment Processing App

Every sale is captured in the payment processing app at checkout, whether paid by cash or card. Typically, payment processing apps have their own reader hardware (as referenced above) and popular providers include Square, Paypal, Clover, and Quickbooks, which is the one we use because it is seamless with our Quickbooks accounting software. The payment processing app functions as a virtual cash register or point-of-sale system (POS) in which the cashier selects the items, enters their weights, then processes the payment. Whichever payment processing app you choose, the simplest way to use it is to assign each cashier a login, then have them download and run the app from their own phones. You could also buy inexpensive tablets to dedicate for this use, just be aware the tablets will need to be connected to the internet for the transactions to go through.

Features to look for in a payment processing app include:

- Ability to run on a variety of smart phones and tablets.
- User-friendly interface with few steps and screens to execute sales and minimal opportunities to make mistakes.
- Integration with your accounting software.

Ticketing App

The ticketing app is the software that pickers will interact with to book spots. We've used several booking and ticketing apps over the years and even considered paying a developer to create one before landing on the one we use today, Ticket Tailor. Any ecommerce platform can be used to sell tickets, but there are features unique to this business that make a ticketing app more practical. Features to look for include:

- A widget element to embed code that will allow customers to book directly on your website. People are leery of following links that take them away from the website they're shopping on to buy something - they are much more comfortable making a purchase directly on the intended website, which the widget enables them to do.
- User-friendly interface with minimal steps and screens to complete the transaction. There's a reason Amazon has a one-click buy button; it sells more stuff. It should be as easy as possible for people to book. A ticketing app I used several years ago required the buyer to enter a name for each ticket holder, greatly slowing down the purchasing process, and our customers *did not like it.*
- One-step refund mechanism that will refund the money, cancel the booking, and reopen the spots for booking by someone else. Unfortunately, I've yet to

find a system that allows the buyer to do the canceling themselves, so it's still a process that requires management from the seller side.

- Ability to collect and export the email addresses and phone numbers of ticket holders by day and by time slot. These can be used to communicate important updates like changes in produce availability or cancellations due to weather.
- Ability to set tickets to automatically open for purchasing at a set time in the future. When we started getting popular and the spots filled up within seconds of opening, Dad once joked that we're the Taylor Swift of farms. Such high demand was thrilling, but the leg work of setting up the booking slots in real time was intensely stressful for me. Picture multiple alarms, me white-knuckled at my laptop logged into the back end of the ticketing app, trying to hit "publish" on the tickets at *exactly the right second* while thousands of people clicked "refresh" over and over on the page. This led to glitches, mistakes, tickets posting inconsistently, and several small strokes. Finding an app that allowed us to set an automatic ticket drop time has solved this problem completely.
- Customizable confirmation emails which can be configured to include additional information like what to bring and wear. Here's our current receipt, which is automatically sent upon purchase. The date and time are populated with the guest's specific booking info through merge tags.

> Thank you for booking spots for the Full Farm Pick on May 29th at 10:00am. Please arrive as close to the beginning of your time slot as possible. It's best to wear closed-toed shoes, hats, and sunscreen to the farm. We also suggest bringing your own water and bags to take your produce home. *You must purchase everything you pick. Produce quantities and availability are not guaranteed.* The farm does our best to target the correct number of pickers for the amount of produce ready to pick, but we cannot always predict how heavily we may be picked before you arrive. The last checkout of the day is thirty minutes after the last time slot ends, and that is when we close completely for the day. We recommend nearby Erwin Park for picnics afterward.
>
> Please determine your route to the farm in advance and ensure you've entered the address correctly, and that your mapping app didn't autopopulate anything! We are located at 7505 County Road 201 in McKinney. From the south we recommend exiting 75 on Laud Howell, not Bloomdale, to avoid road closures from construction.

We will stay open for sprinkles, but not for thunder and lightning. If we need to cancel for bad weather, we'll send you an email and text message the morning of the pick and your admission will be refunded. We will always wait until the morning of to decide on the weather, picks will NOT be automatically canceled for a bad forecast!

If you need to cancel your tickets, simply reply to this email requesting cancellation. Your request must be sent by midnight the night before the pick to receive a refund. *Your admission fees will not be refunded if you no-show, if you email to cancel your spots later than midnight the night before the pick, or if you bring fewer people than you paid for to the farm.*

Take a moment to refer to our Picking Guide [linked] for recommendations on cleaning, storing, and cooking your awesome fresh produce.

Mass Texting App

Text is the quickest and most reliable way to reach customers. Phone numbers are collected during the booking process primarily in case we need to cancel a pick due to weather. Typically, mass texting services, like ClickSend and BulkSMS, charge by the number of recipients and the number of characters in the text. Don't wait to sign up for a mass texting service when it's about to storm and you desperately need to tell a couple hundred people not to head to the farm - buy a small package to learn how to upload numbers and send a test to yourself in advance.

Determine the Layout

Consider the way guests will flow through the experience when they visit. The check-in area should be the closest to the parking lot; if your *checkout* tent is the first people see, that's where they will go, and you'll be redirecting people all day long. Make sure the check-in area is highly visible and use barriers to keep people from wandering straight from the parking lot into the field, or some will.

Scan the QR code to view a video tour of our layout on TikTok.

Set up a spot near but separate from check-in for pickers to gather before you open for the day. It's not like we have a front door to keep locked until opening time – some people *will* arrive early and get out of their cars. If you don't have somewhere to send these folks, they will mill around *precisely* in your way while you're setting up, asking how your day is going or to hear the history of the farm as you frantically count cash boxes and refill coolers and log into apps and check timesheets. We set up a few picnic tables on the parking lot side of the check-in tent, and now it's very easy to say "Welcome! You can wait right over there, and we'll let you know when we're ready to open." This area also functions as a little picnic and gathering area after they check out. Don't make it too comfortable, or you'll have trouble getting people to leave at the end of the pick; memorize directions to the nearest park so you can suggest it to folks wishing for more outdoor time.

Place trash cans only beyond the checkout, on the way out of the farm; if they're accessible from the field, people will fill them with carrot, beet, and onion tops along with anything else they've picked and decided they don't want anymore. And when the other guests see one person cutting tops and tossing stuff, they think they're supposed to do it, and now you've got a mess to clean up. Keeping trash cans located away from the fields and past checkout has been the best remedy.

Try to tuck restrooms close to the action, but out of the way. It is obviously not ideal to position the restrooms in direct line of sight to where most of the activity is going on. Provide a sink for hand washing but post a sign to discourage people from washing their produce on the farm. The main reason for this is that it makes checkout a muddy mess. They never actually remove all the dirt, particularly when it's sticky clay like ours, so instead of weighing dry vegetables with a little dirt on them, the cashier gets stuck handling sloppy wet vegetables covered in mud, which ends up all over their phones, the scales, and the tables. Yuck! If someone wants to wash a few berries or carrots for the ride home after they've been weighed and paid, absolutely, go right ahead.

A little dirt is good for you.

Another point to consider here is the security of the property. For the first few years, we strung a chain across the driveway to prevent people from entering when we were closed. It was quick, cheap, and totally insufficient. As the farm grew in popularity, we added a "closed" sign to the chain, but it didn't stop everyone from coming in. Now we have a proper gate which we open about thirty minutes before the pick begins and lock closed as the last guest leaves.

A gate that locks is invaluable.

Parking

Insufficient parking can be a bottleneck on the whole operation. The parking lot must be large enough to park the number of pickers you can host at the season's peak. If you have a huge, well-draining, perfectly placed field for parking, well, that's great. Our parking area is none of those, but it is sufficient if parked properly.

The most important thing to consider about parking areas is safety. Ensure cars are prevented from driving too close to check-in or checkout areas, where parents will be distracted while their kids amuse themselves. We use logs and traffic cones to create a wide berth around walking-only areas. It's necessary to use physical barriers here; if you don't, someone may try to park right under your tents.

To figure out how many parking spots are needed, translate the number of bookings into the number of spots. This may be easy to glean from the ticketing app, but it's also easy to do quick and dirty on the back of an envelope. Consistently, we sell an average of 2.5-3 spots per booking or party. Assuming that all the spots from one booking are coming from the same household, that makes one car per 2.5 spots sold. We typically top out at 80 spots per hour, so using the conservative 2.5 spots per car, we will need to park 32 cars. Remember that not everyone from the previous time slot will have left before the guests in the next time slot arrive, so when calculating how many parking spots are needed, it will be *more* than the number of guests per time slot. Padding by about a third is safe.

If you're working with a blank open field like we are, it's best to provide some kind of template for people to park properly, by which I mean in straight lines. This will maximize the number of cars that can fit. Our four or five staff members park in a template every morning, setting the ends of each line with our cars for the guests to intuitively park next to. Even so, if you have limited space for parking, plan to have a staff member spend the first 30 minutes of the day waving cars to park in the right spots, keeping the lines of cars spaced appropriately, and preventing any cars from becoming blocked in. If the field is parked correctly at the beginning of the day, it generally stays in good-enough shape for the rest of the pick. Inevitably, if we don't set the template with our cars first thing and assist the first wave of cars, someone will park sideways in the middle of the field and waste a lot of parking spots.

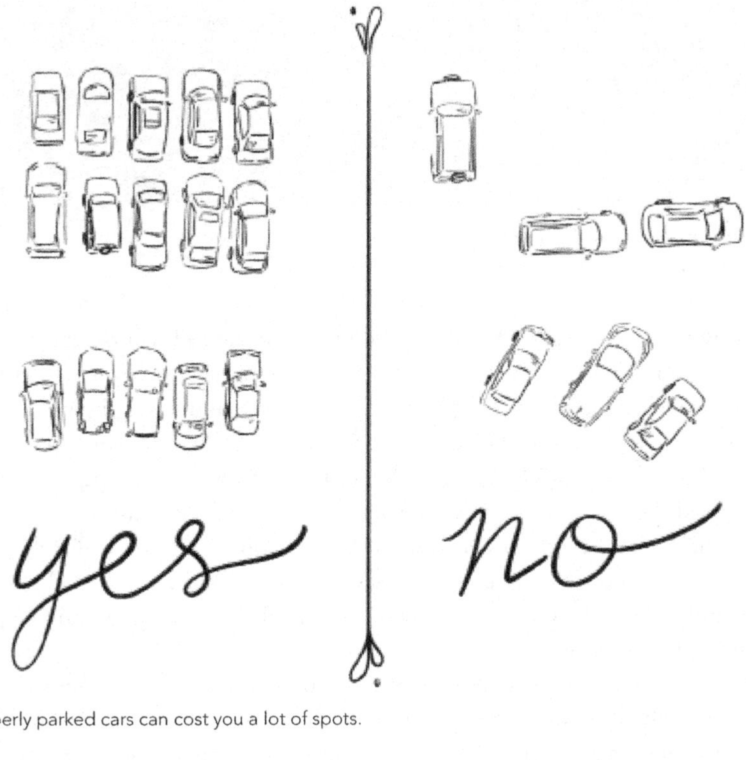

Improperly parked cars can cost you a lot of spots.

In addition to making sure you have enough space for parking, consider how it will hold up after it rains. A couple hundred cars can quickly turn a wet field into a muddy slip 'n slide. Mitigate this with gravel or wood chips where necessary.

Staff

It's very important to us at Pure Land to keep our staff as small as possible. We do not employ any help with farm labor; we choose not to grow more than two full time employees can manage. However, managing the tourism aspect to get everything picked does require some help.

A very slippery parking lot after a few days of rain.

For the first two years, Dad and I also ran the picks all by ourselves. We were growing significantly less back then and only letting in about 40 guests per hour. Dad would walk around the field helping people pick and I was the cashier. The first real gap we noticed was the need for a permanent greeter, someone to welcome guests and get them started - a sign with instructions wasn't cutting it. Then, as we expanded the amount we were growing and increased the number of spots per hour, we needed a second cashier to keep the checkout line under control. At our current scale we operate every pick with one greeter, one floater in the field helping people, and two to three cashiers, totaling about 36 hours of part time help per week through the season.

Think of your staff as ambassadors for your farm. It's vitally important that they are friendly, fun, and fast on their feet. Under no circumstances should you employ anyone that's going to be less than positively chipper when greeting the 239th guest on a hundred-degree day. Yes, I know; these folks are not easy to find.

These are the main qualities you should be looking for in your pick staff:

- Friendly, cheerful, and outgoing
- Patient and comfortable repeating themselves
- Familiar with fresh produce
- Comfortable being outdoors (in our case, in the heat)
- Able to take control, talk loud, and wrangle groups of people
- Comfortable interacting with and respectful of various cultures
- Able to make quick, smart decisions under pressure and direct themselves without you

- Flexible with scheduling, as picks are only scheduled a few days in advance
- Reliable and prompt
- Desire to be part of the family, not just collecting a paycheck
- Bonus: Funny

If you aren't starting off with someone in mind, where are you going to find candidates? Start with your customer base, the people who already know and love your farm. Post to social media and ask your regulars if they're interested or know anyone who might be. As the work is during the summer and only provides relatively low hours, responsible high school or college students may be a good fit.

Our core Pure Land Farm team in 2023.

Take care of your staff!

Take care of your staff. We could not run our business without them, and it's important that they feel like the invaluable members of the team that they are. Encourage staff to pick and take home as much as produce they like and support them when they encounter a difficult guest. I bring cold refreshments for my team every single pick because it brings just a little bit of comfort to a long morning of hustle. Every season starts with donuts and ends with a team dinner.

Set

Before opening for the picking season, establish the farm's procedures, prices, and operational policies. These inform the customer of what to expect when doing business with you. The way your business is conducted should always appear fair and sensible to the public. As the saying goes, if you're explaining, you're losing, so don't let things get overly convoluted. Keep policies simple and intuitive to facilitate a smooth operation where your guests can have a great time while you maximize your profit.

Communicate the following information in as many places as you can: on your website, on the booking confirmation receipt, in social media posts, and on signs at the farm. While ours is not a particularly complicated business - we're just yanking vegetables out of the ground, after all - there are more decisions to make than one might expect. What are you going to charge for admission and each crop? How close to the picking time will you allow someone to cancel their booking and get a refund? What will you do if you see someone picking and dropping produce in the field?

Admission and Booking Strategies

As any business owner knows, one can never have too many customers. It is far better to sell out early at the farmers market due to unexpectedly high traffic than to have produce left over at the end. The same is true at the pick-your-own, but having a huge crowd of pickers at once doesn't provide a particularly great experience, nor is it conducive to maximizing crop yields. It is essential to control customer traffic to provide a quality experience and reach the farm's earning potential. Even with this customer limit in place, there are strategies that can boost the bottom line.

Charge Admission

As cited earlier, a UC Davis study from 2005 reported that 68 percent of respondents were willing to pay between $1 and $15 in entrance fees for an agritourism activity. Only 5 percent indicated they were not willing to pay entrance fees at all. Admission currently makes up more than 25 percent of our revenue, effectively raising all our produce prices by 25 percent, but the guest doesn't experience it that way. Remember, agritourists recognize the value of what they're purchasing is greater than the value of the produce alone.

Deciding how much to charge for admission is like threading a needle. Too little, and you may wind up with lots of tourists just there for a cheap activity, uninterested in the produce that makes up three quarters of the farm's revenue. Charge too much, and people might not come at all. So how much should you charge? It depends on the level of demand for your activity. In our third year of operation, we were stopped on two occasions by guests asking us to raise the cost of admission, hoping that would discourage some of the tourists and make it easier for them to get spots. We raised admission from $2 per guest to $5 the following year and there were no complaints. It may or may not have slowed how quickly the spots fill up but we continue to sell out, which is what matters.

You may encounter guests who ask if they still must pay admission even though they are not going to participate directly in the activity of picking. Our policy is that the admission fee permits the guest's entrance into the farm, whether they are going to be picking or not. The only people excluded from admission are children under the age of one.

Limit Both the Number of People and Party Size

Over time you will get a sense of the number of people you're comfortable hosting at once and the number it will take to pick the field clean. In the beginning, err on the conservative side with fewer people. It sucks to lose produce because there weren't enough pickers, but it sucks *way more* when people spend their time and money to come to the farm and find slim pickings. Nearly every negative review we've ever received has been related to the guest feeling there wasn't enough of a particular crop to pick. This is why we don't open the booking spots more than a few days in advance, to try to hit the right number of people. Even when there is a massive amount of produce ready, though, there is a cap to the number we allow in at one time.

Think about the number of people you let into the farm in two ways. First, obviously, they are individual human bodies that take up space and interact with each other. Controlling the number of bodies on the farm fosters a sense of calm, keeps lines short, and prevents guests from feeling crowded. One day early on, before we started checking in guests by name and confirmed the number of spots they booked, we noticed *way* more people in the field than there should have been. There had to be a hundred children running up down and across the rows, flinging tomatoes and scissors, feeding off each other's frenzied energy like they were on a playground. It turned out that one woman had organized her large "Mom's group" to come out together and incorrectly informed the other moms that they only needed to book their *own* spot, not the children's. It was not a super fun day.

However, the absolute number of people you admit to the farm does not necessarily correlate with the amount of produce that will get picked. The more reliable determinant of how much money you're going to make on any given day is the number of *parties* booked, so think of the number of people you admit to the farm as grouped into *transactions*. Say a party of six guests makes a booking to come pick.

Limiting the number of people on the farm is good for the field and makes for a better experience.

The individual participants may be a mom, dad, three kids, and grandma. When they come up to the register to pay, they will likely be making one transaction, paying for everything they picked all together. If those six guests had been three unrelated parties of two people, however, there would have been *three* transactions at the register, and three transactions from parties of two almost always adds up to more than one transaction by the party of six. For this reason, we limit the number of people per booking to six, large enough to accommodate most families while boosting the number of transactions overall.

Offer a Season Pass

If a product has limited availability, access becomes another product you can sell. We've tried a variety of strategies to increase the number of transactions without significantly increasing the number of people we host, the most effective of which has been a Season Pass program. Season Pass holders may come to any pick without booking a spot, all season long. They still add to the total number of people on site, but they're more likely to be regulars.

Revisiting the types of agritourists discussed in Chapter 2, there are always plenty of tourists that want to visit the farm, but it is better for our bottom line to

prioritize access to regulars. Regulars spend more money with us over the season simply because they visit more often, many every single week as part of their grocery purchases. This is the group for which we actually created the Season Pass in the first place when we noticed our picks selling out too quickly for many of our regulars to get spots. When I reached out to a few to check that everything was alright and invite them to come even if they couldn't get a spot, they would insist they didn't want to be a burden. It turns out, most people don't like feeling like they're sneaking in through the back door and cutting off a bunch of people waiting in line. This was a huge problem - our biggest supporters, the relatively small group of people who buy from us all season, were being edged out by the much larger group of people likely to visit only once or twice. The Season Pass program worked beautifully to correct this. Our regulars have open access to come as often as they like, and we can toggle the number of single-visit tickets for sale down for any given pick as needed to regulate the number of people on site. For example, we sell half the number of single-visit tickets on Saturday as on any other day, because Saturday is by far the most popular among our Season Passers.

2023 Pure Land Farm Season Pass Rates:

- **Individual Season Pass,** admits 1-2 guests: $35
- **Family Season Pass,** admits up to 4 guests: $55

Individual Season Passes are of highest value to us, yielding one transaction at the cost of, at most, two people on site. We added the Family Season Pass later to satisfy guests with more people in their household but limit the number available more strictly to keep the total number of people down.

Start off by selling just a few Season Passes, then scale up year-on-year after observing how they are used. The ticketing company we use, Ticket Tailor, created a feature to track the Season Passes redemptions by scanning a unique QR code sent to each buyer. At the end of the season, we can see how many times each buyer visited the farm. Breakage is a term that describes the revenue gained by retailers for products or services never claimed, the classic example being unredeemed gift cards.

Many businesses love breakage, because they get to pocket the revenue and never have to hand over the goods. It's a little more complicated for us because the number of Season Passes must be kept limited to maintain control over the number of people on the farm. I would much rather sell a Season Pass to someone that's going to use it frequently than to someone who never shows up, which was 24 percent of our Season Pass buyers in 2023. Every spring, we offer the Season Passes in a presale to the previous year's buyers excluding those who did not use their pass the previous summer, hoping to gain a higher concentration of true regulars among pass holders.

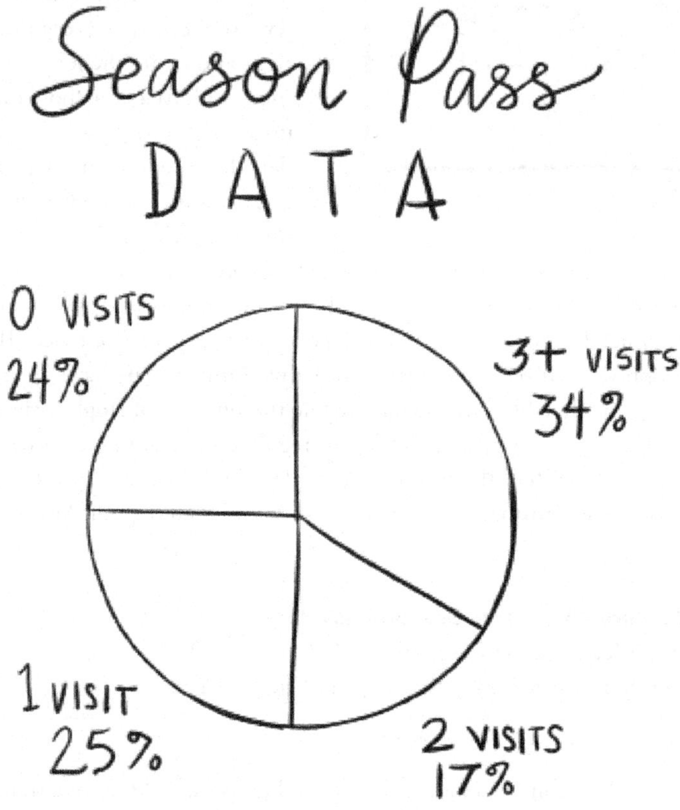

Season Pass redemptions in 2023.

Be Consistent

Consistency is important in an agritourism venture just as it is at the farmers market. One great feature of the pick-your-own model is the flexibility to be open or closed whenever we like within the parameters we've trained customers to expect. Our pick sessions are always three one-hour slots and always held in the mornings. The spots for the weekend, the most popular every week and the ones that sell out

Whoever visits the most wins a gift card to our favorite farm-to-table restaurant, Harvest Season Kitchen.

the fastest, always open for booking on Thursdays at 10am. By staying consistent, the customer doesn't even need to check our website or social media; they always know when spots for the weekend will go on sale.

Pricing Your Produce

Every market is different, so a good starting point for pricing your crops is in alignment with the other growers in your area. Go to the farmers market and see what everyone else is charging for retail. Remember, agritourists aren't motivated by bargains, so there is no need to set prices any lower than retail.

We have toyed with the idea of charging one flat price per pound for all crops as some you-picks do, rather than pricing them individually. If you aren't growing produce with a wide range of price points, this would probably work out financially and save a lot of time at check-out. However, using a flat price per pound for all produce would mean giving up the invaluable yield data captured by the payment processing software, which allows you to keep your finger on the pulse of production to detect problems and hone your growing techniques. Therefore, we don't recommend it.

Post prices in several places before the cash register to avoid sticker shock. Prices should be published online, on a sign at the farm where the guests will see them before they start picking (like where they pick up their baskets), and again on the signs at the end of each row. Because the picking activity is fun and each party may contain several people picking, the resulting bill can be higher than guests are used to at the grocery store. We've learned over the years that most people have no idea what "a pound" of something looks like, so we incorporate visual aids and descriptions as well. For example, "This bucket holds four pounds of blackberries which would cost $28," or "Big tomatoes usually weigh half a pound each." Play around with weighing your produce to get familiar with these measures. Some customers come with a plan to pick a large quantity for putting up, but most are winging it or have four children that each grab their own baskets and start picking wildly, oblivious to how much those eighteen pounds of beets are eventually going to cost. Again, providing information in as many places as possible increases the odds it will be seen or heard.

Occasionally, a guest will come up to the checkout with, let's say, dated expectations about prices based on what the Ma and Pa pick-your-own in their country town used to charge fifty years ago. This guest might say something like, "Wow, that's so expensive! We used to get cucumbers for $2 a bushel!" I bite my tongue to stop from asking how little money they believe farmers should make, as Ma and Pa couldn't possibly have been profitable at those prices, but usually reply something like "Yup, things sure have changed!" Although it doesn't happen often, I like to be prepared with a few standard, polite responses to someone with opinions about our prices. The point is to avoid replying with sass and instead say something that will make them think *oh, that's true - I never thought about that.*

- "Yup, things are more expensive than they used to be! Think about what your house probably cost back then."
- "It's pretty expensive to grow this way, as a small family-owned business. We can't compete with the scale of Big Factory Farms, but our stuff is way better quality."
- "Do you garden?" If they say no - and it's almost always no - you can reply "It's actually pretty hard and takes a lot of work!" If they say yes, you can reply "Then you know how much work it is!"

Operational Policies

This is where you must decide how you're going handle all the various situations that will come up in the normal course of doing business, the nuts and bolts of interacting with your customers. Think of these operational policies as the standard principles of your business, the guidelines you'll use to solve problems and make decisions.

Farm Rules

No one wants to feel hovered over, but you certainly do need to keep an eye on the field while pickers are active. We post a short list of Farm Rules at the check-in tent where parents and children may stop and read them together. Damage is always going to be part of the pick-your-own experience, but it can be minimized with education and supervision. The most important thing to convey before allowing folks to head into the field is to be careful with the plants. Children will try to step or jump across rows anyway, and a friendly "Please be careful!" will usually alert the parent that you have one eye on them. Any children purposefully inflicting damage or picking and throwing produce (yes, it'll happen) will get direct intervention from one of us and be led back to their parent. This kind of thing is a fact of life in any business that serves children.

Post basic rules for picker behavior in a conspicuous spot.

Nevertheless, you may be surprised to learn that children are no worse than adults when it comes to causing damage. Dad practically had to drop his coffee one morning and sprint to stop a full-grown picker who was working her way down a row, uprooting and discarding young cucumber plants. When he asked her what she was doing, she answered, "Trying to find the cucumbers underneath!" I'm not kidding, we have a million of these stories. Of course, you can't lock someone up for this (I know, I called). All you can do is redirect the guest, clean it up, and cuss about it later when everyone's gone home.

Cancellations, No Shows, and Gate Crashers

Life gets in the way of coming out to the farm sometimes, so it's important to have a fair refund policy that gives the guest ample opportunity to cancel their booking before a pick. We administer refunds no questions asked, except when:

- The guest cancels their spots later than midnight the night before the pick
- The guest brings fewer people than they booked to the farm
- The guest doesn't show up

The refund process should be initiated on the day it is requested and usually takes a few days to complete. First thing in the morning before a pick, process any refund requests that came in before midnight so that last minute bookers can snap up the spots (hint: tell pickers you do this, and they'll be on the website waiting for cancelled spots to become available).

The purpose of having guests pay admission ahead of time online is to minimize the number of no-shows. Every party that books and doesn't show up costs at least one transaction, possibly multiple, depending on the makeup of the group. Whole party no-shows are a real bummer, but even partial no-shows are a problem too; booking five spots but only showing up with three people wasted two spots that could have been booked by someone else, which cost the farm one or two transactions. Back in our first year of pick-your-own, we only required the guest to reserve their spots online, then collected the actual payment at the farm. The rationale was to avoid having to manage online payments and refunds, but in reality, it created a greater problem. Regularly, up to a full third of the booked guests did not show up to the pick, a frustrating predicament as we had a field full of vegetables and a deep waiting list of people that wanted to come. Once we started collecting payment online during the booking process, the no-shows virtually stopped.

A picker that cancels their spots five minutes before the time slot opens is as good as a no-show. For others to reasonably be able to book the newly opened spots and make their way to the farm, the spots must become available a few hours before the pick. In the case of last-minute illness, you can offer to the guest that they may come out another time without booking, allowing you to keep their admission and collect on the transaction later. A couple of extra people at any given pick is no big deal, and it happens infrequently enough that we don't bother with a formal rescheduling process.

Set up admission as an item in the payment processing app so you can also take admission payments on site. If someone makes it to checkout and admits they have an extra person they haven't paid admission for, you can just add it to their produce tab. It's also an easy thing to waive, especially for people who are lovely or buying a lot of produce, and it leaves a good impression. It's important to remember most of the farm's revenue will be earned from the produce, not from admission.

Police admission just enough to discourage people from trying to pull one over on you or throwing off the balance of your crop mix, but not so much that you start tagging ears. If a party arrives at the farm with more people than they booked, give them a little lecture, charge them for the extra admissions, and let them in. If the produce is abundant and the farm is calm, a party arriving with no reservation (usually the result of driving by or hearing about us from friends) will get a short primer about how we work, charged admission, and allowed in. But be cautious; if word gets around that you don't truly enforce the booking system, you're going to end up with *a lot* more drop-ins, which is bad both for hitting the right number of pickers for the right amount of produce *and* frustrating for people who play by the rules. If a walk-in drops by and the pickings are getting slim or the farm already feels like a zoo, we'll pull up the FAQ on their phones and politely turn them away. It's better to have walk-ins return on an abundant day than to let them in and give the impression we never have much to pick.

Weather

When there's rain in the forecast, people will start to send messages asking whether the pick is still on. If you decide to cancel preemptively and it doesn't actually rain, you'll be stuck with a field full of vegetables and no pickers. No matter how dire the forecast is, we always wait until the morning of the pick before deciding if we need to cancel. We've been burned by an inaccurate forecast too many times. Of course, it's also not ideal to cancel five minutes before the pick when everyone is on their way. We wait until two hours before the pick starts, and if rain seems inevitable or there is a significant chance of severe weather, we'll cancel. This is communicated by email, social media, and text; not everyone will check their email or social media on Saturday mornings, but everyone will see a text.

If we have to cancel a pick on the weekend, which always sell out fast and for which spots are hard to get, a nice thing to do is allow those folks access to the booking link for the following weekend ahead of everyone else. These small, simple actions can earn you tremendous appreciation from customers.

Opening and Closing Times

Be strict about your opening and closing time for the day but be relaxed on the picker's start times. We allow pickers to arrive at any time during their time slot, which works well to stagger them upon entry and not overload the greeter. But we have a firm close thirty minutes after the end of the last time slot ends, and as the last car pulls out the gate is closed and locked behind them.

Produce Handling

If you pick it, you bought it. This is the one picker behavior that frustrates farmers most. We are considerably less bothered by damage caused by ignorance than by finding produce deliberately picked and then dumped in the field (or as Dad and I call these sad, abandoned vegetables, wounded soldiers). Once while ringing up a regular, she whispered that she witnessed a lady in the cucumber rows tell her children to drop all the normal-sized cucumbers they had picked and brought to her because she only wanted very small ones. When the offender eventually made her way to checkout, I firmly but politely told her I knew what she had done and to please not do it again, explaining that the produce they left in the field had been wasted now that no one else could pick it. She blanched, apologized, and left quickly.

We clean up the field after every pick with a quick walk through the rows to collect everything, then sort wounded soldiers to minimize food waste. Good stuff goes home to our own fridges and is shared with neighbors, damaged stuff goes on the compost pile (of course, we don't expect nor *want* pickers to buy damaged produce). Some days there are only a couple of beets out of place and other days we may collect a couple baskets worth of perfectly good produce. Because we grow a high volume and have diversity of crops, we can comfortably bear the loss. Consider a farm that only grows apples, for example, and because apple trees only make so much fruit, a pile of discarded apples can really take a chunk out of the revenue. I also believe there is a critical mass of production at play; if a farm is very small, any produce lost will obviously sting more than a farm producing a lot. But I can say with absolute conviction that on our scale, the loss we experience through pick-your-own is far less than we've ever had to give away or compost for lack of buyers at the farmers market. Generally, keeping the produce quality high and providing ample picking information are the best defenses against wounded soldiers.

Some people will mutilate vegetables in the field, justifying that they don't want to pay for this part or that because they add weight on the scale at checkout. The aforementioned beet, onion, and carrot tops are the usual victims and we'll often

Teach pickers how to use all edible parts of the vegetables.

find them wrenched off in small, random piles. Aside from the waste, these make a mess. If the tops have been removed before reaching checkout, what's done is done and there's no point bringing it up. But if a picker requests to remove the tops at the register, tell them about some uses for these less-understood ingredients like the recipe for carrot top pesto on your website. You'd be surprised how many people are stunned that you can use the green stalk on top of a fresh onion as… drumroll please… green onion. And throwing away beet greens should be punishable *by law!*

Checkout

I'll go over the basic daily checkout procedures in the next section, but there are a few standard policies regarding what we charge for and what we do not.

Occasionally, a guest will try to abandon some of the contents of their basket at the checkout, declaring that that *physical act of picking* is all they came to do. When this happens, we gesture to the sign at check-in and remind them that they must purchase everything they've picked, information provided to them several times at this point, then suggest recipes, freezing directions, or that they consider sharing the harvest with their neighbors.

A picker with a full basket ready for checkout.

Do not charge for minimally damaged produce and cull anything that has the potential to be unsafe to eat, inedible, or gross. The guest has the ultimate power of selection at the pick-your-own, and the things they deliberately choose to pick and purchase will blow your mind. Four-pound zucchinis, split carrots, bell peppers no bigger than your thumb - pieces of produce you would never *dream* of taking to the farmers market to sell. First, assess whether a damaged item is indeed what the guest wanted to pick. If it's a giant zucchini, you'll know they wanted it because they'll roll up bragging on that sucker, but squishy cherry tomatoes are often tossed into baskets by rule-following parents after being gnawed on by toddlers too. Throw in anything that's minimally dinged but safe to eat for free, letting the guest know you've done so and to eat the item as soon as possible. That's a win-win-win; they've removed a damaged piece of produce from the field, the produce will probably get eaten because it's perfectly edible, and the guest is excited to get something for free. But do not ever let a guest take home damaged produce that might be unsafe, inedible, or bad tasting. Keep a cull bin behind the cash register for bitter yellow cucumbers and squishy melons. If the guest is disappointed to see their pick discarded, offer to help them find a good one back out in the field.

Over six years and thousands of transactions, I can only recall two occasions when someone reached out dissatisfied over the quality of their purchase after getting home. Because it's impossible to control the quality of every single piece of produce, the only prudent thing to do is refund without question and invite them back for another try.

GO!

Show time! Here are the routines we follow before, during, and after every pick.

Before the Pick

Expect some people to be early. Often guests will come from thirty or more miles away and aren't sure how long it'll take to get to the farm, so they'll give themselves extra time and wind up early. Send these folks to the waiting area.

Pre-pick checklist:

- Have employees sign in on the timesheet.
- Log into payment processing and ticketing apps. Connect to the card readers. Ensure that all necessary electronics are fully charged.
- Set up checkout stations with scales, cash boxes and card readers, and bags.
- Scan the field for issues, uncover any plants under shade or netting, rope off any areas to keep guests out of them or drop the ropes to let them in.
- Deploy or modify any signs as needed.
- Stack baskets and picking tools ready to grab at check-in with your greeter.
- Station one employee on parking lot duty fifteen minutes before picking time starts.
- Pound a cup of coffee and eat something with protein.
- Pee.

During the Pick

At opening time, invite guests to the check-in tent. The greeter should give a standard welcome spiel for each party to prepare them for the experience. Here's an example:

> Welcome! What's the last name on your reservation? [check guest in]
>
> Have y'all picked with us before? [yeses may head on in, noes stay]

But first…

> Here's how it works. You can pick up a basket and pair of scissors, then head into the field. You'll see signs at the end of each row that tell you what's planted there, the price, and instructions on how to pick it. If you need help, just ask! When you've picked everything you want, come up to the checkout tent and we'll weigh everything by the crop so you can pay. The restrooms are on the other side of the barn. Please read the farm rules right here, then grab a basket and head on out!
>
> Any questions? Have fun!

The first hour is my favorite. It's quiet, it's cool (likely the coolest it's going to be all day, anyway), and everyone moves a little more slowly. Most of your serious regulars and Season Pass holders will be there at opening time to get the best pick, so this is the perfect opportunity to bond with them. Scarf one of the muffins they brought you and lead them to the plant you found with a couple blushing tomatoes buried in the middle. This time spent taking care of the people who truly appreciate what you do is so enjoyable and well spent.

Circulate in the field to answer questions and help. Dad is our Chatter in Chief. He spends most of every pick talking to guests, telling old timers the history of the farm, answering questions about soil health, and loosening carrots for little kids to pull more easily.

Keep an eye on guests that look like they need to get into the shade or ones with mobility limitations you can help with, especially as the day goes on. Because our farm is terraced down a hillside, we use our utility vehicle to rescue folks we see struggling to get back up to the top.

After about an hour of picking, most people will start moving towards the checkout. Don't rush them; you want to encourage more picking! It usually works out fine without getting overcrowded when some folks stay longer than an hour as the ones leaving after just 30 minutes with fussy toddlers create space.

As they approach checkout, encourage guests to sort their baskets by grouping all the produce from each crop together. If there's not much of a line we don't push this, but jumbled baskets really slow down the checkout process and can create a big line fast. By the time they reach checkout, pickers are hot, tired, and ready to get into the air-conditioned car - the *last* thing they want to do is stand in a line. Provide a few tables under the checkout tent so they can unload and reorganize their baskets while they wait.

The most efficient way to get produce from the basket to the bag is to handle it as few times as possible. Set up your checkout station so the picker places their basket on one side of the scale, and pop open a paper bag on the other side. Unload the basket starting with the heaviest, firmest items (like onions, cucumbers, squash). Load them on the scale by crop, record the weight in the payment processing app, then place them directly into the bag. Proceed from heaviest to lightest and most delicate (like tomatoes, blackberries, flowers), and don't load each bag with more than 10 pounds.

When bagging produce, tomatoes always go on top.

It's important to train your cashiers to distinguish between different kinds of produce not only to ensure you're charging appropriately but so you capture accurate production data. Make sure cashiers can tell the difference between a beet and a radish or between different types of cantaloupes if they have different prices. For example, the small heirloom melons we grow are more expensive than the large hybrid ones. Inspect the produce for quality control as you weigh it, culling when necessary into a hidden bin under the table.

If you get to the bottom of the basket and find a stray jalapeno or something else very small, just toss it in the bag for free. In a long line situation, it's worth more than those thirty cents to process the transaction and move to the next as fast as you can. More than a couple jalapenos or cherry tomatoes though, or if there's no line, weigh and add them.

Honestly, I love working checkout. It's the last impression you're able to make on a guest and an opportunity to turn a just-okay experience into a great one with hospitality. While you weigh and bag produce, engage with the guests about their experience, particularly the kids. *Parents never forget when someone is nice to their kids.* When you discover a particularly impressive vegetable in the basket, ask the pickers which one of them found it, express your amazement, and hold it up for other people nearby to *ooh* and *ahh*. Make a game of asking pickers to guess the exact weight of something and throw it in for free if they guess right (this has only happened twice). I love asking the kids what their favorite thing was to pick and what they're the most excited to eat when they get home.

This guy was so proud of the huge beet he found!

After the Pick

You did it! After locking the gate and putting up the "closed" sign, take a break and have a bite to eat.

Post-pick checklist:

- Walk through the field to collect wounded soldiers, picking tools, and anything left behind by guests (we have an impressive collection of children's hats and sunglasses). Make qualitative observations as you go regarding what was picked clean and what was not.
- Run the sales report to see harvest yields for the day.
- Count the cash box and deposit in the bank. Get change if necessary to reset the box.
- Put away checkout stations. Set credit card readers, tablets, any other elec-

tronics to charge.

- Wash and restack baskets and tools.
- Update the website and post on social media.
- Do the farm work.

A few sad tomatoes abandoned under a plant.

Troubleshooting

There's too much produce ready to harvest that isn't getting picked.

First, consider improvements to the picking experience that can increase the size of each transaction - in other words, how can you get the folks that are already there to pick more? Mow any grass that's gotten out of control, prune the big leaves that are obscuring the zucchini, and peel back the plastic mulch to make the onions more visible.

If the spots are selling out and the picking experience is as good as it can be, always add more people before adding more days. A field picked by 300 people every third day is picked far *more efficiently* than a field picked by 100 people for three days in a row. Identify what is holding you back from adding more people at one time and try to address it; do you need more parking, another staff member to supervise the field, another cashier?

There's way too much of *one* crop.

With even the most optimized crop mix you can experience an unexpected bumper crop, and that's never a bad thing! As mentioned earlier, additional sales avenues should be considered a last resort because they take you away from the core business, but sometimes it's the best option. Run cleanup and sell wholesale, donate, make value-added goods, or compost the surplus.

There's not enough produce ready to make it to the end of a pick.

Size down the number of spots or spread out the picking days. Some crops can be picked every day, but others need time for the fruit to grow or ripen in between pickers. To maximize revenue, most of what is picked needs to be of an appropriate, mature size. Picked too aggressively, these crops will never reach their revenue

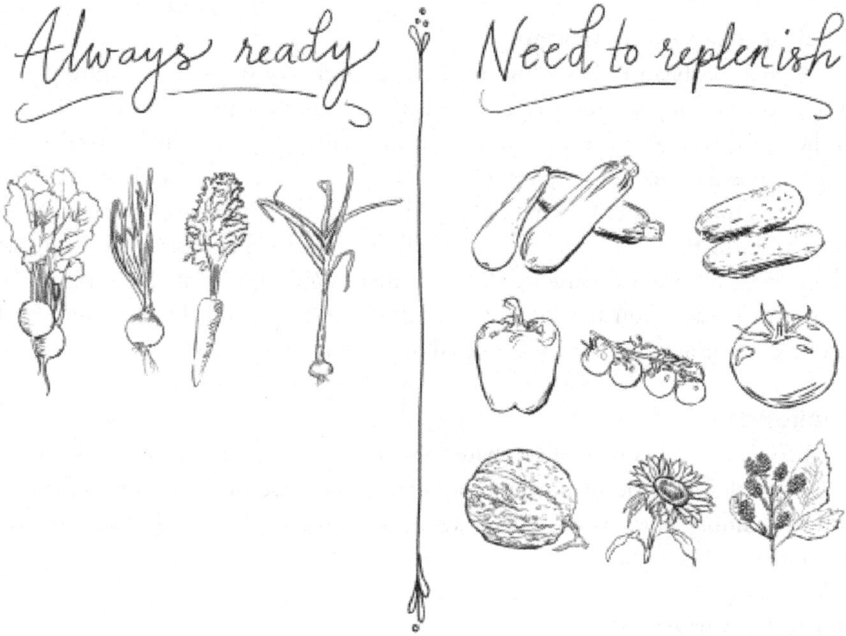

Some crops can be picked every day; some need a break.

potential because people will *pick them down to nubs* if allowed access. Sometimes it can take weeks for the plants to recover enough to bear a significant yield. Build in time between picking days for the plants to replenish and have signs ready to close rows if necessary (a physical barrier like rope works better, but still not perfectly).

Not enough people are signing up.

Go back to Chapter 5 and work to build your following. Post on social media regularly, infiltrate those mom groups, cross promote with other local farms and businesses, pitch your story to local press, and encourage your customers to spread the word. Targeted social media ads can also be a cheap and effective way to boost your numbers. Don't slack off on those weekly emails.

Another reason could be heat, which puts a real damper on our demand as the summer goes on, despite having a substantial social media following and email list. Start the first time slot as early as you can - we start at 8am, but I'm positive there would be takers at 7am. If your earliest slot sells out but no one signs up for the later ones, add as many spots to that early time slot as you can and figure out what you need to do to manage that many people at once, like adding a cashier.

Lastly, take a good hard look at the quality of your produce and the guest experience. Word of mouth, whether personal or via social media, is a huge driver. Are your customers enthusiastic? If not, why not?

Too many people want to sign up.

A good problem to have, but one that can feel very stressful. Be upfront and tell customers that you simply cannot meet the demand you're experiencing. Utilize all the tools available to remove yourself from having to personally be involved as much as possible: set auto-replies to email and social media, pin an informational post to the top with a link to the FAQ, and set spots to automatically open consistently with the ticketing app. Encourage folks to join the email list to get the inside scoop. It has been our experience that customers highly appreciate the certainty of a confirmed reservation and the quality of the experience with a limited number of people at the farm, even if they struggled to get that reservation.

Challenges

Farmers are no strangers to challenges. The very nature of farm work entails long hours, hard physical labor, and the willingness to accept an enormous amount of risk. Running a pick-your-own creates some additional flavors of anxiety to prepare for.

Stress Management

The hardest part of this entire endeavor might be managing the mental stress of the season. It's a very intense eight weeks, and we live and breathe our business the whole time without breaks. A reporter doing a piece on the farm once asked me, incredulously, "You guys are here *every day*, Monday through Friday?!" and I replied, "… and Saturday and Sunday." But it's doable, because it's finite. While we *love* what we do, we also start the countdown clock on opening day.

As a self-diagnosed "people pleaser," I feel a personal responsibility to ensure our guests have a good time. This is good because it makes the business better, but it's important to establish some emotional separation, too. We have poured our hearts and souls into this little patch of earth and really, really want people to like it. Here are some techniques we use to manage our stress.

- Set aside times that you can completely turn off, even just for a couple of hours. During the season, it isn't possible to do this for very long. Hundreds of people are planning their schedules around a visit to the farm and they have questions that need answering or to cancel, so checking messages and managing tickets has to be kept up with multiple times a day. At the farm itself, there *is always* something that needs doing, from weeding to reapplying Bt to mowing to irrigating to fertilizing to fixing the sign that blew away. But when you feel like you're drowning, just do what you *must*, then close the laptop, lock the gate, and walk away for a while.
- Build in redundancies and safeguards that make it hard to forget things or make mistakes. First, identify what you would really be screwed if you didn't have. Have backup methods to take payment if the payment processing app is down, like a second processing app or Venmo. Bring portable charging

bricks to keep phones and card readers full of juice. Stock extra batteries for the scales. Buy bug spray and sunscreen in multiples. Keep any items you need to take to the farm with you by the door with your keys.

- Remove other sources of stress from your life, whether temporarily or permanently. Go to bed early and eat as well as you can. If you find socializing draining after entertaining a couple hundred people a day, don't do it - my friends know I generally do not accept invitations during the picking season. At home, clearly communicate what you need from your partner and family to manage your stress level. At the farm, surround yourself with a team that has fun, works hard, and is as dedicated to the farm as you are.
- Do as much as you can to get ahead before the picking season starts so that you'll be able to focus on the unforeseen crop, pest, and people issues as they inevitably pop up. For example, laying landscape fabric between the rows for weed control costs money and takes a lot of pre-season time, but is far preferable to needing to mow or cultivate between rows for guests when the spider mites show up. If you don't have enough time to address both the weeds *and* the mites, things can go downhill fast.
- Don't lose perspective. *They're just vegetables.* We're not saving lives here. Whatever happens, the customers will get over it.

Keeping Produce Quality High

Remember, the number one thing people like about coming to the farm is getting high-quality produce. Apologizing to pickers for cracked beets and split tomatoes gets old very fast. Either figure out how to improve the quality of a problem crop or don't grow it; pickers might not even miss it. As you gain growing experience, the quality of your produce should improve every year. But inevitably, by the end of June, things start looking rough around here. It can be really tough to keep the quality high the last week or two of the season to eke out that last little bit of revenue, but, bottom line, we end our season when we are no longer consistently proud of our produce.

Unrealistic Expectations

Because every year is a little different, we are constantly tweaking our crop mix to supply the right amount of ripe produce to the right number of visitors every day. We once received a one-star review from a picker who came especially for cherry tomatoes fairly early in their season and was upset to find none left by the time they arrived for their time slot, the last of the day. The review asserted we should have *emailed or texted everyone mid-pick* to tell them there were no more cherry tomatoes left. Maybe one could do this on a single-crop farm like strawberries, but it is virtually impossible on a diversified operation like ours. I tell the story to illustrate the very high expectations some people will have of your little vegetable farm. Just do your best to hit the right crop mix, cover your bases by communicating what's going to be widely available or limited ahead of time, apologize if something is out, and if all else fails, blame Mother Nature.

Managing Heat and Severe Weather

The increasingly severe weather we're experiencing here in North Texas is hard on plants, people, and infrastructure. During the 2023 picking season, extreme storms destroyed our tents *twice*, marking the second and third times that's happened in six years. It was far too hot without shade, so we had to cancel full picks to put the tents back together, creating a domino effect of produce going overripe and needing to be cleaned up and hauled out. Add to this the stress of managing refunds and consoling the couple hundred guests that were expecting to visit, and it made for a very not-fun time. In 2022, we had the opposite of severe storms - there was not *one drop* of rain the entire picking season and temps were in the triple digits every day, which did a number on both the produce and the demand. *You can't fix these problems.* Work around the weather as best you can, but don't sweat what you can't control.

Severe weather not only damages infrastructure but canceled picks can cause the field to get ahead of the pickers, resulting in waste and extra labor on cleanup.

Chapter 7
OTHER AGRITOURISM ACTIVITIES

Pick-your-own is only one of many types of agritourism activities taking place on farms all over the world. If your farm is already set up for visitors, what other kinds of agritourism programming could you offer? Think about reasonable, complementary activities; the simplest and most cost-effective way to expand your enterprise is to work with what you already have and what you already know. In other words, don't invest in a bunch of equipment to host cheesemaking classes if you aren't already making cheese.

Before adding a new agritourism activity to your roster, consider the following:

- How much of your effort will managing the activity take from booking through execution? Must it require *your* involvement, or can it be assigned to an employee?
- Is the activity seasonal? If so, does that season overlap with times you are already pretty busy?
- What investment is required up front and what will the ongoing costs be?
- Is there competition providing similar activities? How will you distinguish yourself?
- How much can you charge and how many participants can you realistically host? In other words, project the potential revenue this activity will bring in.

Classes

At Pure Land, the only supplemental agritourism activity we offer are educational classes we call Field Trips. We hold these in April and early May before the picking season starts, as the farm schedule is too busy and unpredictable once picking gets underway. The income isn't terribly significant, usually only 3-4 percent of our annual revenue, but we enjoy doing them and they're great advertising directly to our target audience. Nearly everyone who comes out for a Field Trip says they'll come back during the picking season, and they often do. I used to lead these myself, a strain on my time during the critical spring growing period, but my excellent greeter now handles them all for us.

Our greeter Teri leading a Field Trip.

Everyone starts a seedling to take home.

Here's a sample one-hour Field Trip suitable for all ages:

> Welcome! Here are the farm rules to keep you safe.
>
> Does anyone here have a garden? What do you grow? What's your favorite? Discuss and try to get everyone involved.
>
> These are our farm's regenerative growing practices. Tailor the level of detail to the age group.
>
> This is our farm's hardiness zone and the planting calendar we use. What could we plant right now? Show posters.
>
> These are the parts of a plant and the annual plant life cycle. Show demo transplant and poster. Distribute seeds in jars for everyone to look at and guess what they grow. Compare seeds of the same plant families.
>
> What are the differences between fruits and vegetables? Hold up magnets of fruits and vegetables to guess and stick them under the correct label on a whiteboard.
>
> There are good bugs on the farm and bad bugs. Show poster. Distribute bug boxes.
>
> Walk around the farm through all the crops. Hunt for bugs and identify any caught. Talk about each crop.
>
> Write names on pots, fill with soil, and bury seeds to

take home.

Any questions? Come back at picking time this summer!

Materials:

- Laminated posters: Map of US hardiness zones, planting calendar, parts of a plant, beneficial insects
- 1 demo transplant
- 12 jars containing different kinds of seeds
- Assorted fruit and vegetable magnets
- Magnetic whiteboard
- 25 bug boxes
- 2" planting pots
- Sharpies
- Potting soil
- Picnic tables for the participants
- 6' table for the Field Trip leader
- Bin to hold all the above materials

A natural place to start is with schools. If you'd like to start hosting schools on field trips, approach them at the district level. There can be a lot of hoops to jump through in getting set up to work with schools, but generally once you're in, you're in. For a school to consider visiting your farm you'll likely need to align the program with the educational requirements of the district, which can usually be found online. Each grade level has distinct knowledge touch points listed for each school subject, which you can use to tailor your program. For example, in Texas, all First Graders are required to be able to identify and compare the parts of a plant, so to make our program appealing to the school district, we include learning the parts of a plant. Comb through the educational requirements in your state, identifying the ones that can be met on the farm for each grade level, then plan your program around them.

A local Goddard School Summer Camp on a Field Trip.

Schools aren't the only folks that

may be interested in educational farm classes. Preschools are often looking for this kind of educational activity for their students, as are many youth organizations. To start hosting Girl Scouts, Boy Scouts, and the like, research the requirements to earn any relevant farm, science, or outdoorsy merit badges and pitch your programs to their local councils. Another group we've found to be particularly interested in Field Trips are homeschoolers. More and more families are opting out of public school but don't want to forgo the social aspect entirely, so they form groups of like-minded families. Searching with Google and social media are good ways to locate these groups and reach out. Finally, we also offer a few open Field Trips where families may book as individuals and are grouped together in mixed sessions. Those are always a lot of fun.

Farm Tour guest Christie enjoying a cocktail at Pure Land.

Educational farm classes for adults can also be quite popular. Every year we host a few late-afternoon Field Trips for Grown Ups. The program is essentially the same as the kid's Field Trips, but BYO. A McK-

Guests enjoying a walk through the crops.

inney nonprofit, the Seed Project Foundation, holds a fundraising Farm Tour every spring, where a hundred guests board buses to tour three different local farms, enjoying cocktails and snacks along the way. A friend of mine owns a flower farm and holds flower arranging classes with fresh blooms in the summer and dried ones in the winter, enabling her to host them year-round. Her classes are especially popular with bridal parties where she guides the bride and bridesmaids through creating their own bouquets for the big day.

Tips for classes:

- Set a minimum number of participants. If a staff member is leading the activity, ensure that minimum revenue more than covers the cost of their labor. If a group doesn't have enough kids to reach the minimum, we allow them to book at the full price for the minimum number if they wish. A maximum number of participants also helps to keep the quality of the experience high. We've found that the wheels start falling off once you exceed about 25 children in one group. If you have a larger group interested in coming out, consider breaking the field trip up into stations and rotating smaller groups through.
- Set a minimum age. We charge for age 3 and up. Younger kids can tag along but really aren't able to participate.
- Set the available time slots with the same ticketing app you use for the picking season. Don't allow groups to suggest the times at which they want to come. Having Field Trips all over the place will ruin your schedule.
- Require payment up front and in full through the booking process. Do not get bogged down collecting payment on site right before the activity starts, waiting for late and no-show participants and ending up with fewer than your minimum. If it's a group, let the group's organizer collect payment from their participants. Let the group's organizer also handle any refunds from participants who can't attend, as long as they continue to meet the minimum number of attendees.
- Whenever possible, schedule multiple sessions back-to-back. If you're getting all the materials out and scheduling staff to host a Field Trip, it's more efficient to do two or three in a row.
- In the booking process, make it very, very, *very* clear what the activity entails. Despite specifying several times that our Field Trips take place before the picking season starts and therefore *do not include picking*, invariably at least one person every year expresses disappointment that there was no picking. And for some reason, many moms want to know precisely every single thing the kids will be doing, so we've taken to publishing a detailed outline of the activities on the booking page.
- Aim for a mix of lesson-style teaching, Q&A, and hands-on activities to keep participants engaged. Kids get antsy very quickly if they're not actively

involved in some way. Don't overestimate their attention spans, particularly under the age of 6.

- *Be fun.* They're kids! But they will *run all over you* if you let them. Take charge with a loud, commanding presence (a friendly one, obviously) and don't let things get out of control.

Other Agritourism Activities

Perhaps you have an attractive old barn, a lake loaded with fish, or a commercial kitchen on site. The possibilities are almost endless for adding activities that could supplement your farm income, a particularly appealing prospect during the non-picking season. Depending on your resources and facilities, you could consider branching into:

- Weddings and special events
- Farm stays
- Camping
- Farm dinners
- Hunting, fishing, and hiking
- Photography sessions
- Cooking classes
- Birthday parties

An Outstanding in the Field dinner at Pure Land in 2017.

Chapter 8

GO FORTH

Now, it's your turn to open the gate and let people in.

A few months back, I stopped at one of the farmers markets where we used to sell when we first started growing. I recognized almost no one; hardly any of the farmers that were market gardening back when we began are in business anymore. Frankly, if it weren't for pick-your-own, we wouldn't be in business anymore, either.

When we began our journey into agritourism, creating a community of local agriculture superfans was not really what we set out to do. It just happened as we spent time in the sun with our guests, teaching them about plants, sharing favorite recipes, listening to childhood stories of time spent in grandparents' gardens. What we actually set out to do was make a living growing food, and I still believe that's the critical foundation upon which other good things become possible. Agritourism made Pure Land Farm profitable, yes. But while the dollars we earn only benefit *us*, agritourism helps *all* farmers by raising awareness of regenerative local farming, making personal connections between the public and the agriculture community, and fostering intentional consumers. No, we didn't set out to win hearts and minds, but people just handed them over, and we were kinda blown away. Imagine the possible impact from small farms practicing agritourism *everywhere* - a surge of profitable farmers both strengthening their local food economies and exponentially growing that vital base of conscientious consumers we need to affect real change.

There is desperate need for more small farmers to regeneratively restore our lands and build robust local food systems, and there are many eager hopefuls who would love to pursue it as a career, but we cannot continue pretending the current model is working. Dr. Booker T. Whatley pointed this out forty years ago. To add to his sentiments (from my own soapbox), it's not okay to let young people give up precious years of their youth pursuing market farming when we know damn well most of them won't be able to make a living from it. Likewise, we should not feel great about farmers markets full of retirees that are hobby farming to stay active, only able to return week after week because they don't need to make any money. *That's not a sustainable local food system.*

We're losing farmers to suicide and prime farmland to developments every day. Does anyone care? I think a better question is, does anyone *know?* Does the general public know how important a robust local food system is? Have most of them ever set foot on a working farm? As farmers, it's pointless to complain that the public won't pay the still too-low prices we usually ask for our goods. Why won't they pay more? Because they haven't the *foggiest idea* how we got that food onto the table. They don't know because no one has taught them. Municipalities and developments continue to unabashedly devour open space, impeding or prohibiting the production of local food because most of the public *does not know* that supporting the local food movement is critically important to building vibrant, healthy communities and protecting our environment.

We growers must extend our hands to the public and do the hard, necessary work of educating them about the value of local, regenerative agriculture. I truly believe the most effective way to build support for these causes is through the power of fun, immersive experiences on farms. Everybody likes to have fun, and feeding ourselves from the earth is a deeply ingrained, lizard brain impulse - tapping into it has a great capability to move people. Indeed, we have observed that the public is not apathetic at all about these issues when they become aware of them at our farm. That's the heart of agritourism, to peel back the curtain on something unfamiliar, *farming*, and invite people to make it part of their lives.

A final word. It's easy to feel very small and overwhelmed working outside under the great wide Texas sky, but never underestimate the impact you have as a person that cares for a patch of earth. The opportunity to steward a piece of land, however big or small, is an awesome gift and responsibility. There is so much value in what you do, in some ways that can be quantified monetarily but in many more that cannot. Not everyone sees that value yet, but we will show them, one by one. I hope this book enables *you* to start showing people the value of what you do, too.

One morning a couple years back, I thought we had checked out the last picker of the day and started to pack up but noticed one car left in the parking lot. One of our regulars was deep down in the field reclined on a grassy slope, her basket of vegetables in the sun beside her, just quietly taking it all in. We waved and went back to work, letting her be until she was ready to face the world again.

We wish you the best on your journey into agritourism. May you be amazed at the community that gathers around you. Go forth and share the pure joy of growing things.

Published by Acres U.S.A. Books

Acres U.S.A. is North America's oldest publisher on production-scale organic and sustainable farming. For more than fifty years, our mission has been to help farmers, ranchers and market gardeners grow food profitably, regeneratively and without harmful chemicals.

We offer a full range of educational products—including books and pamphlets, a monthly magazine, webinars, podcasts, online courses and both large-venue and on-farm events—to help producers find timely, practical and well-tested knowledge and advice on just about any agricultural topic

Acres U.S.A. is committed to providing information about agricultural production systems that can regenerate the soil, support the farmer and rancher, and bring healthy, nutritious food to the consumer. We bring both new and time-tested voices to the discussion of "what is to be done" when it comes to building soil health, raising healthy animals and growing nutritious crops. We amplify the work of modern agricultural leaders and innovators, and we support all efforts to increase the resilience of our farms and ranches—and the people who anchor and feed our nation.

The books we sell, and the speakers at our events, and the editors of our monthly eNewsletter and farmer-facing website represent generations of experience in building—not mining—soils. The lessons they teach are fundamental: to farm economically, one must do so ecologically.

Visit acresusa.com

to learn about our magazine, events, books, and other educational opportunities.

Other titles from Acres U.S.A. books:

The Biological Farmer

By Gary Zimmer

Biological farmers work with nature, feeding soil life, balancing soil minerals and tilling soils with a purpose. The methods they apply involve a unique system of beliefs, observations and guidelines that result in increased production and profit. This practical how-to guide explains their methods and will help you make farming profitable and fun. Biological farming does not mean less production; it means eliminating obstacles to healthy, efficient production. Once the chemical, physical and biological properties of the soil are in balance, you can expect optimal outputs, even in bad years. Biological farming improves the environment, reduces erosion, reduces disease and insect problems, and alters weed pressure and it accomplishes this by working in harmony with nature.

Farming Without Losing Your Hat

By Paul Dorrance

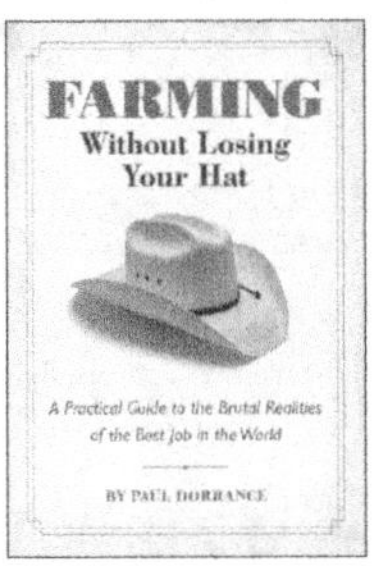

If you're looking for an honest, straight-forward, no-holds-barred book about the challenges of building a successful farming business, then Farming Without Losing Your Hat is the book for you.

Author Paul Dorrance shares from his own real-world experiences, taking a practical-only approach, focusing on the less-glamorous but essential parts of farming and ranching successfully today.

Additionally, space is provided at the end of each chapter for you to write down and capture your starting vision and goals. By the time you finish the final page, this book will have transcended into an heirloom book capturing where you were when you first dreamed of the farm and how you planned to make it a reality. Keep it forever in your farm's library, and pass it along in your family for generations to come..

Holding the Lines

By Maureen Ash

Farmer's daughter Maureen agreed completely with her mother's admonition: Never marry a farmer. Her marriage to a science geek with a good job meant a house on a lake and clean hands-until the farmer hidden inside him started scrabbling to escape. What began as a willingness to put up with his buying a tractor and renting a field becomes an all- encompassing endeavor to learn how to drive, raise, train, and work draft horses, grow and sell their crops, care for the land they had learned to love, maintain a sense of humor, and keep themselves and their children safe in the face of runaway teams, straight-line winds, and sub-zero temperatures. Holding the Lines is the story of a very ordinary woman's decision to find meaning in horses, hard work, love, and potatoes.

My Farmer, My Customer

By Marty Travis

This book belongs in the hands of anyone who eats and wants to understand what it takes to produce healthy food. Through new ideas and a different perspective, Marty shares a winning formula for success in modern family farming.

Written by the co-owner of the Spence Farm featured in the documentary, "Sustainable," this book provides the reader with the necessary steps an owner-operator must take to set up their family farm for long-term success. It's a must read for anyone who is passionate about healthy food and healthy farms, from chefs to home cooks to the commercial grower.

Marty, a seasoned farmer, writes each chapter like a neighbor giving trusted advice, including honest details about what worked and what failed — more reasons this book is so special.

Made in the USA
Monee, IL
13 September 2026

62976956R00098